T0178711

Network Coding

Network Coding

Edited by
Khaldoun Al Agha

First published 2012 in Great Britain and the United States by ISTE Ltd and John Wiley & Sons, Inc.

ISTE Ltd
27-37 St George's Road
London SW19 4EU
UK

www.iste.co.uk

John Wiley & Sons, Inc.
111 River Street
Hoboken, NJ 07030
USA

www.wiley.com

© ISTE Ltd 2012

Library of Congress Cataloging-in-Publication Data

Network coding / edited by Khaldoun Al Agha.
    p. cm. -- (ISTE ; 618)
Includes bibliographical references and index.
  ISBN 978-1-84821-353-1 (hardback)
  1. Coding theory. 2. Data transmission systems. 3. Computer networks--Mathematical models. I. Al Agha, Khaldoun.
  TK5102.92.N47 2012
  004.6--dc23
                                                                                        2012005892

British Library Cataloguing-in-Publication Data
A CIP record for this book is available from the British Library
ISBN: 978-1-84821-353-1

Printed and bound in Great Britain by CPI Group (UK) Ltd., Croydon, Surrey CR0 4YY

Table of Contents

Chapter 1. Network Coding: From Theory to Practice 1
Youghourta BENFATTOUM, Steven MARTIN and
Khaldoun AL AGHA

   1.1. Introduction 1
   1.2. Theoretical approach 2
      1.2.1. Max-flow min-cut 4
      1.2.2. Admissible code 5
      1.2.3. Linear code 6
      1.2.4. Algebraic resolution 6
      1.2.5. Random code 8
   1.3. Practical approach 10
      1.3.1. Topologies 11
         1.3.1.1. Multihop wireless networks 11
         1.3.1.2. Cellular networks 18
      1.3.2. Applications 19
         1.3.2.1. Network coding and TCP 19
         1.3.2.2. Network coding and P2P 21
         1.3.2.3. Network coding with priority 22
   1.4. Conclusion 23
   1.5. Bibliography 24

Chapter 2. Fountain Codes and Network Coding for WSNs

Chapter 2. Fountain Codes and Network Coding for WSNs 27

Anya APAVATJRUT, Claire GOURSAUD,
Katia JAFFRÈS-RUNSER and Jean-Marie GORCE

2.1. Introduction   27
2.2. Fountain codes   29
   2.2.1. Generalities   30
   2.2.2. Families of fountain codes   33
      2.2.2.1. Random fountain codes   33
      2.2.2.2. Luby Transform (LT)   34
      2.2.2.3. Raptor code   40
      2.2.2.4. Code complexity   41
2.3. Fountain codes in WSNs   41
   2.3.1. Implementation   42
   2.3.2. Protocol of reliability enhancement:
       ARQs versus fountain codes   43
   2.3.3. Discharge and overflow   45
2.4. Fountain codes and network code for
     sensor networks   49
   2.4.1. Impact of network coding on
       the degree distribution of an LT flow   50
      2.4.1.1. XOR network coding and LT code   50
   2.4.2. Design a network code for LT code   54
      2.4.2.1. Solutions of network coding   55
   2.4.3. Application to multihop sensor
       networks   58
      2.4.3.1. Multihop linear networks   58
      2.4.3.2. Sensor networks   61
2.5. Conclusion   66
2.6. Bibliography   67

**Chapter 3. Switched Code for *Ad Hoc* Networks:
Optimizing the Diffusion by Using Network
Coding**   73
Nour KADI and Khaldoun AL AGHA

3.1. Abstract   73
3.2. Introduction   74
3.3. Diffusion in *ad hoc* networks   77
3.4. Diffusion and network coding   78
3.5. Switched code: incorporate erasure codes
      with network coding   83
    3.5.1. Definitions   84
    3.5.2. Coding function of switched code   84
3.6. Decoding function of switched code   85
3.7. Design and analysis of a new distribution   87
    3.7.1. Analysis of switched distribution   90
3.8. Conclusion   96
3.9. Bibliography   97

Chapter 4. Security by Network Coding   99
Katia JAFFRÈS-RUNSER and Cédric LAURADOUX

4.1. Introduction   99
4.2. Attack models   100
    4.2.1. A type-II wiretap network   102
    4.2.2. A nice but curious attacker   104
4.3. Security for a *wiretap network*   105
4.4. Algebraic security criteria   106
    4.4.1. Note on random linear network coding . . .   107
    4.4.2. Algebraic security   109
    4.4.3. The algebraic security criterion   109
    4.4.4. Algorithmic application of the
            criterion   111
4.5. Conclusion   112
4.6. Bibliography   112

Chapter 5. Security for Network Coding 115
Marine MINIER, Yuanyuan ZHANG and Wassim ZNAÏDI

5.1. Introduction 115
5.2. Attack models 116
   5.2.1. Eavesdroppers 117
     5.2.1.1. Internal eavesdroppers 117
     5.2.1.2. External eavesdroppers 117
   5.2.2. Active attackers 118
     5.2.2.1. Pollution attacks 118
     5.2.2.2. Flooding attack 119
   5.2.3. Definition of homomorphic ciphering
      schemes 120
     5.2.3.1. Two specific schemes 122
     5.2.3.2. Completely homomorphic
       encryption schemes 123
   5.2.4. Homomorphic encryption and
      confidentiality in network coding 124
     5.2.4.1. The case of network coding
       using XOR 125
     5.2.4.2. The case of network coding
       in general 127
5.3. Confidentiality 128
   5.3.1. Alternatives for confidentiality 128
5.4. Integrity and authenticity solutions 130
   5.4.1. Definitions of homomorphic MAC and
      homomorphic hash functions 132
     5.4.1.1. Definition 132
     5.4.1.2. Examples of such schemes 133
   5.4.2. Definition of homomorphic signature
      schemes 134
     5.4.2.1. Definition 134
     5.4.2.2. Examples of such schemes 135
   5.4.3. Alternatives for integrity and
      authenticity 136
     5.4.3.1. Polynomial method 137

      5.4.3.2. Method using checksums   139
      5.4.3.3. Overlapping MAC   140
  5.5. Conclusion   142
  5.6. Bibliography   143

Chapter 6. Random Network Coding and Matroids   147

Maximilien GADOULEAU

  6.1. Protocols for non-coherent communication . . .   148
     6.1.1. Routing   148
     6.1.2. Random linear network coding   149
     6.1.3. Random affine network coding   151
     6.1.4. Example and comparison   152
  6.2. Transmission model based on flats
      of matroid   153
     6.2.1. Matroids   153
     6.2.2. Model and comments   156
     6.2.3. Matroids for SAF, RLNC, and RANC   158
  6.3. Parameters for errorless communication   160
     6.3.1. Rate, delay and throughput   161
     6.3.2. Number of independent elements
        received   164
  6.4. Error-correcting codes for matroids   167
     6.4.1. Operator channel and lattice distance . . .   168
     6.4.2. Matroid codes   170
     6.4.3. Matroid codes for SAF   171
  6.5. Matroid codes for network coding   173
     6.5.1. Rank metric codes   173
     6.5.2. Matroid codes for RLNC   175
     6.5.3. Matroid codes for RANC   177
  6.6. Conclusion   180
  6.7. Bibliography   181

Chapter 7. Joint Network-Channel Coding for the Semi-Orthogonal MARC: Theoretical Bounds and Practical Design

Chapter 7. Joint Network-Channel Coding for the Semi-Orthogonal MARC: Theoretical Bounds and Practical Design 185
Atoosa HATEFI, Antoine O. BERTHET and
Raphaël VISOZ

    7.1. Introduction 185
        7.1.1. Related work 186
        7.1.2. Contribution 188
        7.1.3. Chapter outline 190
        7.1.4. Notation 191
    7.2. System model 191
    7.3. Information-theoretic analysis 195
        7.3.1. Outage analysis of SOMARC/JNCC 196
        7.3.2. Outage analysis of SOMARC/SNCC 200
        7.3.3. Types of input distributions 202
        7.3.3.1. Gaussian i.i.d. inputs 202
        7.3.3.2. Discrete i.i.d. inputs 202
        7.3.4. Information outage probability
                achieving codebooks 203
    7.4. Joint network channel coding
        and decoding 203
        7.4.1. Coding at the sources 203
        7.4.2. Relaying function 204
        7.4.2.1. Relay detection and decoding 205
        7.4.2.2. JNCC 206
        7.4.3. JNCD at the destination 208
        7.4.3.1. SISO MAP detector and demapper . . . 208
        7.4.3.2. Message-passing schedule 209
    7.5. Separate network channel coding and
        decoding 212
    7.6. Numerical results 214
        7.6.1. Information-theoretic comparison of the
                protocols 215
        7.6.1.1. Individual ϵ-outage capacity with
                Gaussian inputs 215

7.6.1.2. Individual information outage
probability with discrete inputs   216

7.6.2.   Performance of practical code design   219

7.6.2.1. Comparison of JNCC functions:
XOR versus general scheme   220

7.6.2.2. Gap to outage limits   222

7.6.2.3. Comparison of the
different protocols   224

7.7.   Conclusion   226

7.8.   Appendix. MAC outage performance
of high SNR   228

7.9.   Bibliography   230

Chapter 8. Robust Network Coding   235

Lana IWAZA, Marco Di RENZO and Michel KIEFFER

8.1.   Coherent network error-correction codes   237

8.2.   Codes for noncoherent networks,
random codes   240

8.3.   Codes for noncoherent networks,
subspace codes   242

8.3.1.   Principle of subspace codes   242

8.3.2.   Recent developments   244

8.4.   Joint network–channel coding/decoding   245

8.4.1.   Principle   247

8.4.2.   Recent developments   248

8.5.   Joint source–network coding/decoding   249

8.5.1.   Exploiting redundancy to combat loss . . .   250

8.5.1.1. Artificially introduced correlation   250

8.5.1.2. Existing correlation   253

8.5.2.   Joint source–network coding   254

8.6.   Conclusion   256

8.7.   Acknowledgments   257

8.8.   Bibliography   257

Chapter 9. Flow Models and Optimization for Network Coding 265

Eric GOURDIN and Jeremiah EDWARDS

9.1. Introduction 265
9.2. Some reminders on flow problems in graphs . . 267
9.3. Flow models for multicast traffic 272
9.4. Flow models for network coding 277
9.5. Conclusion 284
9.6. Bibliography 285

List of Authors 289

Index 291

Chapter 1

Network Coding: From Theory to Practice

1.1. Introduction

Traditionally in a network, the intermediate nodes intervening in the routing of the information flow between a source node and a destination node transmit data packets without processing them. Network coding is an approach that allows the intermediate nodes to mix or combine the packets between them to improve the network performances. These performances depend on the number of packets to be combined, on the way they are combined with each other, and on the information the node must have on its network topology and state to perform the right coding.

Network coding was initially a purely theoretical approach because it required that the coding scenario be centralized and that the existing flows were known in advance. Its evolution

Chapter written by Youghourta BENFATTOUM, Steven MARTIN and Khaldoun AL AGHA.

has enabled us to use random parameters and thus to make it distributed and applicable in practice.

Network coding improves the network performances in terms of throughput because its use makes it possible to reach the maximum capacity of the network, i.e. *max-flow min-cut*. Furthermore, it is robust to losses. In fact, the loss of a packet has less impact on the performances because the information it contains can be recovered from other combined packets. It is also scalable because it is decentralized and uses local information for the collaboration between nodes. Finally, it enables an increased security because a combined packet is only readable provided that all the packets used to make the combination are available.

All these advantages allow network coding to be used in various network topologies (wired, cellular, *ad hoc*, etc.) and for different applications (peer-to-peer, TCP, applications with quality of service (QoS), etc.).

In this chapter, we explain the theoretical aspect of network coding, its principle and the advantage of using it, and then we give examples of its practical applications.

1.2. Theoretical approach

In information theory, the concept of network coding was introduced for the first time by [AHL 00]. The main idea behind this concept is that instead of considering the node as a simple information flow relay, network coding allows the node to combine several packets into a single packet. The history of network coding is presented in detail in [NET 00].

We explain the working of network coding with the example of the wired *butterfly* network illustrated in Figure 1.1.

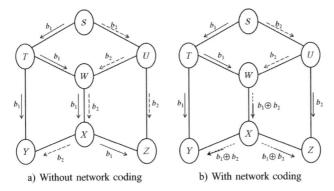

a) Without network coding   b) With network coding

Figure 1.1. *A butterfly network with a multicast flow from S to Y and Z [AHL 00]*

Source node S wants to send two bits b_1 and b_2 to destinations Y and Z (multicast). For this, S sends b_1 to T and b_2 to U. Then, T sends b_1 to W and Y. Similarly, U sends b_2 to W and Z. In a conventional scenario, i.e. without network coding, and by assuming that all the links have a unit capacity, W sends b_1 then b_2 (or vice versa) to X. Then, X sends b_1 to Z and b_2 to Y.

By using network coding, W performs a combination of two bits, an exclusive OR, for instance, and sends it to X. Then, X transmits this combination to Y and Z. Node Y (respectively Z) receives b_1 and $b_1 \oplus b_2$ (respectively, b_2 and $b_1 \oplus b_2$) and can consequently recover b_2 (respectively b_1) by performing the operation $b_1 \oplus (b_1 \oplus b_2)$ (respectively, $b_2 \oplus (b_1 \oplus b_2)$).

We can easily infer that the conventional approach requires ten transmissions, while the approach using network coding requires nine. The coding gain is then $10/9$ in the wired network.

If we consider a wireless *butterfly* network, each transmission is being received by all the neighbors of the emitting node, and the gain would be $8/6$. Here we notice that network coding has more significant impact in a wireless environment.

This section thus explains the theoretical aspects of network coding and its evolution, from its advantage in terms of maximum capacity to its use of random values allowing it to be scalable and put into practice.

1.2.1. *Max-flow min-cut*

Generally, the quantity of information circulating in a communication network cannot exceed a given value, which is set by the maximum capacity reachable in this network. This capacity is defined by the *max-flow min-cut* principle. The use of network coding enables us to reach this maximum capacity.

In graph theory, the *max-flow min-cut* problem consists of finding a maximum flow that can be carried out from a source toward a destination. In a network, this amounts to finding the paths that the data sent by the source must follow to have the maximum flow capacity. For instance, in the *butterfly* network illustrated in Figure 1.1, by assuming unit capacity links and by considering source S and destination Y, S can send at most two pieces of data to Y via $S \rightarrow T \rightarrow Y$ and $S \rightarrow U \rightarrow W \rightarrow X \rightarrow Y$. In this case, the capacity as defined in *max-flow min-cut* between S and Y has a value of 2 (equivalent to the number of independent paths linking the source to its destination in the case of unit capacities). Similarly, the *max-flow min-cut* capacity between S and Z, independent of the flow from S to Y, is equal to 2.

Let us now assume that source S wants to send two pieces of data to two destinations Y and Z. The maximum reachable capacity is equal to the smallest of the maximum capacities for each of the destinations taken separately. This capacity is thus equal to 2. However, without network coding, the unit capacity of the $W \rightarrow X$ link does not enable us to simultaneously route the data of the multicast flow coming from nodes T and U. Thus the maximum reachable capacity of S toward its destinations cannot be reached, except by doubling the capacity of the $W \rightarrow X$ link.

On the other hand, the use of network coding enables the sending of a data combination on the $W \rightarrow X$ link of unit capacity, making it possible to simultaneously route data toward the destinations and hence to reach the maximum capacity equal to 2.

In what follows, we are interested in the feasibility of a communication scenario implementing network coding and enabling us to reach the *max-flow min-cut* bound.

1.2.2. *Admissible code*

In [AHL 00], the authors introduce the concept of network coding and give a characterization of the admissible coding rate.

The capacity of the graph links is said to be *admissible* if and only if the information rate from the source is smaller than or equal to the *max-flow min-cut* value obtained with this capacity. A coding scheme fulfilling an *admissible* capacity is said to be an *admissible* code. Moreover, an α-code is represented by a transaction sequence that describes a communication scenario. Each transaction transfers a piece of information (data combination previously received) from the sender to one or several receivers. An α-code is said to be α-*admissible* if this transaction scenario exists.

The main contribution of [AHL 00] is to prove that the set of *admissible* codes corresponds to the set of α-*admissible* codes. In other words, for networks wishing for traffic with a capacity lower than or equal to that of *max-flow min-cut*, there exists an α-*admissible* code, using in particular network coding, which is associated there with. The proof is done for a source node in a cyclic or acyclic graph.

For multiple sources, if two information sources are independent, the solution is obtained by solving each problem individually (superposition principle). However, the solution is not optimal, in general. We see different coding techniques in the following section.

1.2.3. *Linear code*

There exist diverse methods to combine the packets. It has been shown in [LI 03] that for multicast traffic, the linear code multicasts (LCM) are sufficient to reach the limits of maximum capacity. The use of an LCM makes the coding/decoding process easier and faster to implement in practice [YEU 06].

A linear combination is a sum of packets weighted by coefficients. The assignment of coefficients can be carried out in several ways. For instance, a greedy algorithm has been proposed in [LI 03] for the code construction. The following sections present less complex coding techniques.

1.2.4. *Algebraic resolution*

The coding of packets to be combined allowing us to reach the maximum capacity is not unique. The authors of [KOE 03] propose to solve the network coding problem algebraically. They give necessary and sufficient conditions under which a set of connections is feasible in a network. In the multicast case, a connection transfers the data (stochastic process) from the source node to several destination nodes. In Figure 1.2, for example, node a sends three data units to node d. The transfer matrix having X as input and Z as output is given by $M = A \cdot (I - F)^{-1} \cdot B^T$, where the three matrices A, F, and B are defined as follows:

– A: matrix performing the transfer of input processes to the intermediate processes,

– F: adjacency matrix of the graph in which each vertex is a link, concerning the intermediate nodes,

– B: matrix performing the transfer of intermediate processes to output processes.

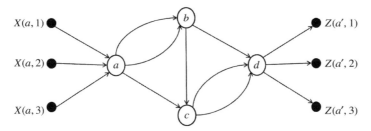

Figure 1.2. *A network representing the random processes [KOE 03]*

The set of output processes $Z = X \cdot M$ depends on the set of input processes as well as on the transfer matrix. In the case of a single source and single destination, the latter recovers the original information by inverting matrix M. The matrix elements are chosen after resolution of an algebraic system, which amounts to finding an algebraic variety[1] that assigns values to elements while keeping a non-zero determinant.

For a network with one source and several destinations, a connection reaches its maximum rate if and only if the *max-flow min-cut* is reached for each connection relative to the source and to one destination. As a consequence, each destination has knowledge of only part of the matrix, the part

---

1 Algebraic variety: set of solutions of a system of polynomial equations.

that concerns it. The authors give a greedy algorithm to solve the system and to show its correctness.

In the case of several sources and several destinations, the system can be solved if and only if the submatrices of the diagonal of M are invertible and the rest of the matrix is equal to zero. In other words, there is no interference between two pieces of information sent to two destinations.

As we can see in Figure 1.1, the coding/decoding scheme must be agreed beforehand and requires full knowledge of the network topology. This constraint makes the implementation of network coding difficult in real cases. In practice, the networks have several constraints: the packets are subjected to random delays and losses and centralized knowledge is difficult to obtain. Hence the solutions must be simple with a low complexity.

1.2.5. *Random code*

To overcome the aforementioned constraints of the networks, linear random network coding has been proposed in [HO 03]. This random coding enables, in particular, a distributed transmission and a stronger robustness to losses. This approach *randomly* extracts the coefficients of the combinations from a Galois field[2]. The coding process is characterized by the generation size and the base field. The generation size is the set of packets used to create the combinations. The base field is the basis of the vector space used to generate the coefficients of the different combinations.

The data to be sent are divided into several blocks of k packets referred to as generations of length k. The packets of

---

2 A Galois field is a finite field, i.e. it has a finite number of elements. For instance, in a domain based on modulo 2^3, we have (6, 4, 1) + (5, 6, 3) = (3, 2, 4).

each generation are combined to generate a new information block. A redundancy can be added to overcome the packet loss. In the scenario presented in Figure 1.3, two encoded packets are lost but the receiver, thanks to the redundancy, manages to infer the initial data after receiving six combinations.

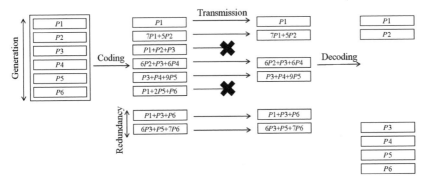

Figure 1.3. *Coding / decoding phase*

Regarding the basis of the Galois field in Figure 1.3, the first two packets, for instance, are generated by using vectors (1,0,0,0,0,0) and (7,5,0,0,0,0). The basis must be sufficiently large to guarantee linear independence between the combinations.

By considering this hypothesis, redundancy can be added at any point of the network to compensate for the lost packets. Furthermore, to determine the packets to be combined, only local knowledge is required. In fact, it is not compulsory that the nodes know the entirety of the transfer matrix. For the decoding phase, each node receives packets that are linear combinations of source packets and saves them in a matrix. Then it must simply solve the algebraic system to infer the original packets.

Network coding has several advantages, such as increasing the throughput and making the network more robust to losses.

For this reason, it has diverse applications in wired peer-to-peer network (P2P), sensor, *ad hoc* and cellular networks [FRA 06].

In the following section, we present some implementations of network coding, mainly in wireless networks.

1.3. Practical approach

Network coding is used in diverse network topologies: wireless multihop networks (mesh and *ad hoc*), cellular networks, and wired networks [FRA 06, NAR 08]. Even though the first theoretical works on network coding dealt with wired networks, it appears in practice that it has more significant impact on wireless networks. Besides, as illustrated by the example of the *butterfly* network in Figure 1.1, network coding is more interesting in a wireless network and in particular in multihop networks. This explains the fact why the majority of works on network coding deal with this type of network. In [FRA 06], the authors have proposed decentralized algorithms enabling us to reach minimum cost multicast connections in wireless networks.

In this section we see some examples of network coding implementation at the level of the routing layer on different topologies: probabilistic and random network coding applied to multihop wireless networks and network coding applied to cellular networks.

We also see its use for other applications involving other layers: P2P application, TCP application, and application with QoS.

It must be noted that network coding is also applied at the level of the physical layer. This variant that is referred to as physical layer network coding (PLNC) is somewhat different from traditional network coding that deals only with discrete symbols (not signals). This approach is described in detail in the following chapters.

1.3.1. *Topologies*

1.3.1.1. *Multihop wireless networks*

1.3.1.1.1. Probabilistic network coding

In this family of approaches, we see mainly COPE protocol and its variants [OMI 08, LE 10, DON 10] as well as an optimal algorithm, ROCX, using linear programming by considering a particular case of the COPE scheme, that of Alice and Bob.

COPE: It is the first architecture of network coding in a wireless mesh network [KAT 06]. It uses network coding for both unicast and multicast flows. It allows the node to combine (binary XOR) several received packets into one. However, the intermediate nodes handle only the native packets (non-coded). As a consequence, they must decode all the packets they receive.

We talk about probabilistic network coding because for the coding, the node calculates the probability that the packets it wants to combine are decoded by the receiver. If it is greater than a certain threshold (0.8), the node sends a packet combination, otherwise it sends a native packet. To calculate this probability, the node needs to know the delivery probability of the links that it must use for the transmission and the packets held by the neighboring nodes. For that purpose, each node sends information contained in the packet header to update the knowledge of its neighbors. COPE adds a header between the frame and the IP headers. It allows the node to:

– give the identifiers of the combined packets in the data field,

– send the acknowledgments (ACKs) of decoded data packets,

– send its reception report (the native packets it has knowledge of).

Figure 1.4 shows that the node can have several possibilities of combinations. Node B, for instance, must send packets $P1, P2, P3$, and $P4$ to nodes A, B, C, and D, respectively. It knows that nodes A, C, and D have knowledge of $[P3, P4], [P1, P4]$, and $[P1, P3]$, respectively. Node B has diverse possibilities to combine the packets, and it must choose the possibility that maximizes the number of packets delivered in a single transmission, here, $P1 \oplus P2 \oplus P3$. In fact, this combination allows the three neighboring nodes to decode.

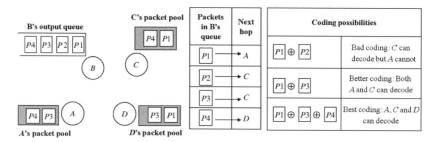

Figure 1.4. *COPE coding scheme [KAT 06]*

The implementation in an *ad hoc* network shows that the use of COPE can increase the flow by a factor of 3 (respectively 1.33) for UDP (respectively TCP) flows. COPE is sensitive to flow symmetry: the more symmetrical the flows are, i.e. the throughput of a flow from a source S to a destination D is closer to the throughput of the flow from D to S, the more efficient the coding is.

Thanks to its input, COPE is one of the most well-known protocols in the area of applicative network coding. In addition, several works have been devoted to its improvement by proposing diverse variants.

BFLY: COPE imposes that each intermediate node decodes the packets it receives before re-encoding and transmitting them. As a consequence, no coded packet is transferred directly. The BFLY approach was proposed in [OMI 08]. The authors noticed that it was possible to improve the gain in a *butterfly* network if the transfer of a coded packet (without decoding it) was possible.

In fact, in the wireless network represented in Figure 1.5, node A (respectively B) wants to send packet $P1$ (respectively $P2$) to nodes D and C.

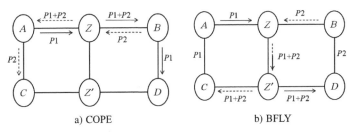

a) COPE   b) BFLY

Figure 1.5. *Coding scheme using COPE and BFLY [OMI 08]*

We notice that COPE requires five transmissions while BFLY requires four. We recall that when node Z, for instance, broadcasts packet $P1 \oplus P2$ via the wireless links $Z \to A$ and $Z \to B$, we consider a single transmission.

The authors propose to combine BFLY and COPE by using a probabilistic model to characterize the reachable throughput. The throughput depends on the delivery probabilities of the links used. For that purpose, the nodes evaluate if network coding using BFLY is more interesting in terms of delivery probability; otherwise it tests if COPE is more interesting; if need be, the packet is sent native (non-coded).

DCAR: [LE 10] extends the COPE scheme by considering two sources separated by more than two hops, each having a destination, which is a neighbor of the other source, and a node in common as indicated in Figure 1.6. During the routing, the source indicates the list of its neighboring nodes, the intermediate nodes then check if they have flows that have followed the paths matching DCAR scheme and then carry out the coding of the packets accordingly.

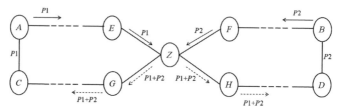

Figure 1.6. *DECAR coding scheme*

COPE with delay: In contrast to the variants based on the improvement of the COPE coding scenario, the approach presented in [DON 10] modifies the algorithm of selection of the packets to be combined. In fact, the need to integrate QoS in terms of delay to each packet is associated with a deadline. Thus, during the selection of the packets to be combined, those having the closest deadline are coded in priority.

ROCX: In previous works, the routing decision is made independent of the coding phase. The authors of [NI 06] improve the gain of network coding by taking into account the coding opportunities during the routing process. The ROCX (routing with opportunistically coded exchanges) algorithm is applied for mesh networks and uses linear programming to introduce the coding constraints while minimizing the total number of transmissions. This approach is particularly efficient when the paths are long but has some drawbacks. Linear programming has a high complexity and cannot be applied in a network where the nodes have a weak computational power. It does not allow scalability because the complexity of the linear program increases exponentially as a function of the number of nodes. Also, each node must have knowledge of the links used for all traffic going through the network, yet this condition is difficult to fulfill in practice. Another drawback of ROCX is that it is based only on the Alice and Bob scheme (shown in Figure 1.7), which is a particular case of coding schemes that can be carried out by COPE. In this scenario, there is a flow from Alice toward Bob and another from Bob toward Alice. When the intermediate node (relay) receives packet $P1$ from Alice, and packet $P2$ from Bob, it combines them and broadcasts the combination $P1 \oplus P2$, then Alice (respectively Bob) decodes and recovers the original packet $P2$ (respectively $P1$).

There is no opportunistic listening in this scenario. In other words, a node considers only the packets that are destined for it even if it manages to receive other packets in the network. ROCX, therefore, cannot exploit the coding opportunities of another scheme such as that of BFLY [OMI 08], for instance. It has been shown that opportunistic listening can be exploited to increase the throughput [KAT 06].

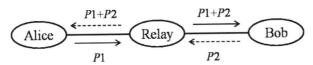

Figure 1.7. *Network coding applied to the Alice and Bob scenario*

1.3.1.1.2. Random network coding

To illustrate this family of approaches, we explain the MORE protocol. ExOR is a protocol that uses opportunistic routing, but does not apply network coding; MORE improves it by integrating the concept of network coding.

Extremely opportunistic routing (ExOR): The traditional routing for the unicast traffic consists of transferring a packet through the network hop-by-hop, from a single node toward a single node. [BIS 05] describes ExOR as a unicast routing technique used to increase the throughput in wireless multihop networks. It is important to recall that ExOR does not apply network coding, but, nevertheless, improves the network performances. By using ExOR, the source node divides the data into blocks (referred to as generations or batches), then broadcasts the packets by specifying a priority between forwarders (the one-hop neighbors responsible for the data transfer). If the forwarder with the highest priority receives a packet, it waits for a period of time, which is a function of its priority and, then sends an ACK. Otherwise, a forwarder with less priority is responsible of acknowledging the packet. As a consequence, a forwarder must only broadcast the packets that are not ACKed by nodes of a higher priority. For instance, in Figure 1.8(a), node B has a higher priority than node C.

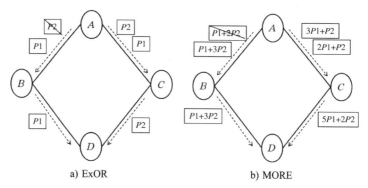

Figure 1.8. *Transmission scenario using ExOR and MORE*

MORE: ExOR increases the throughput but prevents the simultaneous transmissions by using a local scheduling. However, it is MAC dependent, prevents the spatial reuse, and is not applicable to multicast traffic. Another opportunistic routing, i.e. MORE, has been suggested in [CHA 07]. It introduces network coding to overcome these drawbacks.

When a forwarder receives a packet, it checks whether it is innovative or not. A combination is said to be innovative if it is linearly independent from all the combinations previously received, i.e. it carries a new piece of information. If this is the case, the forwarder builds a random linear combination of all the received packets that belong to the current generation and then sends the packet. When the medium is not available, MORE exploits this time to code the packets so that the coded packets are sent immediately when the medium is available. When the destination receives enough combinations to be decoded, it sends an ACK that triggers the transmission of the following generation and suppresses all the packets of the acknowledged generation from the memories of the intermediate nodes.

One of the main differences between ExOR and MORE is that ExOR requires a scheduling to elaborate a ranking among the forwarders. For instance, in Figure 1.8(a), the source must indicate that forwarder B has a higher priority than forwarder C. Another difference is the necessity to have a coordination between the transmitting nodes. Forwarder C must wait to infer that B has correctly received $P1$ but not $P2$. MORE does not require coordination. Nodes B and C send their respective combinations to D, which decodes and recovers the initial data.

Probabilistic versus random network coding: Here we compare the algorithms that are the most representative of probabilistic and random network coding, i.e. COPE and MORE. COPE uses *inter-flow* network coding because it combines packets from several sources, whereas MORE uses *intra-flow* network coding since it combines only packets coming from the same source and more precisely corresponding to a given generation. MORE has a better time management because it codes when the medium is not available. COPE requires more control information (reception report and ACK of each neighbor). The decoding process of COPE is easier because it consists of XOR operation, while MORE calculates the inverse matrix by using a Gaussian elimination. This then concludes that the two approaches have advantages and drawbacks, but are applicable in different contexts.

1.3.1.2. *Cellular networks*

Thanks to its several advantages, network coding has been proposed for cellular networks and in particular distributed antenna systems.

In [CHE 06], the authors prove that by using network coding, the most robust system with respect to losses uses fewer material resources (ASsistant antenna, AS) and enables a more efficient spectral exploitation. In fact, in Figure 1.9, we can see that when the system does not use network coding, each node requires an AS to replicate the data. The use of network coding decreases the number of ASs and hence the number of transmissions. As a consequence, the system uses less bandwidth.

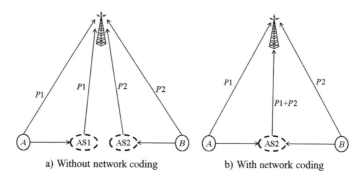

a) Without network coding   b) With network coding

Figure 1.9. *Transmission scenario in DAS [CHE 06]*

In the case where the nodes can cooperate with each other and where there is no AS, network coding enables a greater robustness for the system. One considers that there is a system outage when the base station cannot recover the data sent by a station. Figure 1.10 shows that for a fair cooperation between all the users, each user, after having received its native packet, has three independent paths to transmit the redundant information, whereas it has only two in a network without network coding: without network coding, A, B, and C send $[P1; P2], [P2; P3]$, and $[P3; P1]$, respectively. As a consequence, packet $P1$, for instance, can be transmitted via two paths. However, when network coding is used, A, B, and C send, for instance, $[P1; P2 \oplus P3], [P2; P1 \oplus 3 \cdot P3]$, and $[P3; P1 \oplus 2 \cdot P2]$, respectively. In this case, packet $P1$ can be sent via three paths and thus be decoded. We can easily see that if the three packets are lost, contrary to the traditional system, network coding guarantees the recovery of the three packets and avoids a data loss no matter which packet is lost.

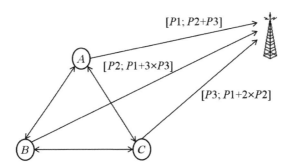

Figure 1.10. *Cooperation with network coding [CHE 06]*

1.3.2. *Applications*

1.3.2.1. *Network coding and TCP*

Many approaches use network coding to improve the throughput at the level of the network layer. However, its application is not restricted to this layer; it also applies to other layers. For instance, [SUN 09a] presents TCP-NC, an improvement of TCP integrating network coding. This approach consists of adding an intermediate layer between the third and fourth layers to combine the segments. The main idea is to acknowledge the degrees of freedom (the number of linearly independent combinations received by the destination) instead of acknowledging the decoded packets. As a consequence, the number of broadcasts decreases significantly in a network having lossy links.

Random network coding is chosen to generate independent generations provided the basis of the Galois field is sufficiently large. Instead of sending a native packet, the source node generates a linear combination of segments lying in its coding window and sends it. The redundancy is used to overcome the losses on the links.

As soon as a combination is received, the destination node saves it in a matrix and uses Gaussian elimination, on the one hand, to find the degree of freedom to be acknowledged and, on the other hand, to decode the initial piece of data when it allows it. By assuming independent combinations, each combination is associated with a degree of freedom to acknowledge. This amounts to finding in each combination the smallest native packet that still has not been acknowledged (*seen*) even if it has not yet been decoded. This technique integrates redundancy because it can become impossible to recover the piece of data before end of transmission of all the combined packets.

In the example illustrated in Figure 1.11, the source sends four combinations and two of them are lost. When the destination receives the first packet, it considers that packet $P1$ is *seen* and hence acknowledges it even if it has not yet been decoded. Then, during reception of the second encoded packet, as $P1$ has already been *seen*, $P2$ is hence the next packet to be considered as *seen* and hence to acknowledge.

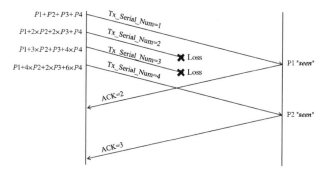

Figure 1.11. *A coding example with acknowledgments [SUN 09a]*

The simulations show that in the absence of losses, TCP-NC can coexist with TCP without affecting their fairness in resource sharing. When there are losses on links, TCP-NC is more efficient than TCP. A variant of TCP-NC, i.e. TCP-NCint, allowing the intermediate nodes to re-encode the data, appeared to be very efficient [SUN 09a].

1.3.2.2. *Network coding and P2P*

Network coding also applies to the level of the application layer. For instance, the authors of [GKA 05] describe a practical system based on network coding for file distribution to a large of cooperative users over a P2P. The advantage of network coding use is that the nodes make decisions on the packets propagation based on only local information. In fact, in Figure 1.12, for instance, A downloads two packets $P1$ and $P2$ and C downloads $P1$. B wants to download $P1$ and $P2$. Without network coding and without any knowledge of transfers in the rest of the network, A cannot make a decision on the appropriate packet(s) ($P1$ or $P2$) it must send to B. By using random network coding, node A sends a combination of $P1$ and $P2$ to B. On the other hand, B receives $P1$ from C and performs the decoding to recover $P2$. These transmissions do not require the nodes to know how the data propagate in the network.

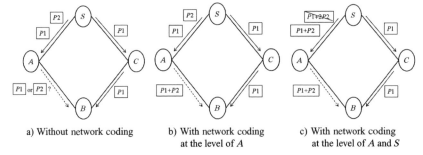

a) Without network coding   b) With network coding   c) With network coding
                                       at the level of A at the level of A and S

Figure 1.12. *A P2P network with nodes having only local information*

The simulations show that the use of network coding improves the downloading rates [GKA 05]. Furthermore, the system is more robust to the loss of servers and of peers containing the copies of data. In fact, if these suddenly disappear, the nodes can recover the file. For instance, by considering the topology of Figure 1.12(c), we assume that B wants to download a file from source S. Node S uses network coding, it divides the file into blocks, and sends combinations to intermediate nodes A and C. By assuming linear independence between the different combinations (basis of the Galois field is large enough), if the server disappears from the system before A has the entirety of the file parts, B can recover the file from the combinations received by A and C.

1.3.2.3. *Network coding with priority*

In general, network coding considers that the packets have the same priority. The authors of [SIL 07] describe a system that combines the idea of priority encoding transmission (PET) with the network coding technique. For certain applications, such as video streaming, the packets do not have the same priority because the first packet of a sequence must be decoded before the following packet.

PET approach has already been used in traditional networks (without network coding). It guarantees the recovery of the packets by complying with their priorities [SUN 09b]. The problem tackled by the use of PET jointly with network coding is that a contaminated packet (erroneous packet introduced by a malicious user) can make the correction of packets with higher priorities that are combined with this packet impossible. This problem cannot be solved with traditional codes used in networks that do not use network coding.

To make the network more robust against losses and corruption (erroneous packets) while using network coding, the approach described in [SIL 07] uses Gabidulin's codes [GAB 85] that are more sophisticated than Hamming's codes[3]. The aim is to guarantee the data recovery even when these codes are used jointly with network coding. The simulations on a media streaming scenario show that the performances are better and that the delay is acceptable compared to the scenario without network coding.

1.4. Conclusion

In this chapter, we have first tackled network coding using the theoretical approach. It presents many advantages because of which network coding has been suggested to be applied to real networks. It has first been shown that the capacity obtained with the use of network coding could achieve the maximum reachable capacity. Then algorithms have been

---

3 A Hamming code is a linear code for error correction. It can detect and correct bit errors by relying on Hamming's distance (number of bits that differ from one code word to another).

proposed to find the appropriate coding technique. Finally, it has been shown that random linear coding has enabled the application of network coding in real networks.

Through some examples, we have illustrated the implementations of network coding in different types of networks, i.e. the multihop wireless networks as well as the cellular networks. Network coding can be integrated at different levels, routing, transport (TCP) and application (P2P, QoS). For each of these implementations, the simulation results have shown the efficiency of network coding.

1.5. Bibliography

[AHL 00] AHLSWEDE R., CAI N., LI S.-Y.R., YEUNG R.W., "Network information flow", *IEEE Transactions on Information Theory*, vol. 46, no. 4, pp. 1204–1216, 2000.

[BIS 05] BISWAS S., MORRIS R., "ExOR: Opportunistic Multi-Hop Routing for Wireless Networks", *ACM Conference of the Special Interest Group on Data Communication (SIGCOMM'05)*, Philadelphia, PA, USA, pp. 69–74, 2005.

[CHA 07] CHACHULSKI S., JENNINGS M., KATTI S., KATABI D., "Trading structure for randomness in wireless opportunistic routing", *ACM Conference of the Special Interest Group on Data Communication (SIGCOMM'07)*, Kyoto, Japan, pp. 169–180, 2007.

[CHE 06] CHEN Y., KISHORE S., LI J.T., "Wireless diversity through network coding", *IEEE Wireless Communications and Networking Conference (WCNC'06)*, Las Vegas, USA, pp. 1681–1686, 2006.

[DON 10] DONG Z., ZHAN C., XU Y., "Delay Aware Broadcast Scheduling in Wireless Networks using Network Coding", *Second International Conference on Networks Security, Wireless Communications and Trusted Computing (NSWCTC'10)*, Wuhan, China, pp. 214–217, 2010.

[FRA 06] FRAGOULI C., LE BOUDEC J.Y., WIDMER J., "Network coding: an instant primer", *ACM SIGCOMM Computer Communication Review*, vol. 36, no. 1, pp. 63–68, 2006.

[GAB 85] GABIDULIN E.M., "Theory of Codes with Maximum Rank Distance", *Problems of Information Transmission*, vol. 21, pp. 1–12, 1985.

[GKA 05] GKANTSIDIS C., RODRIGUEZ P., "Network Coding for Large Scale Content Distribution", *IEEE International Conference on Computer Communications (INFOCOM'05)*, vol. 4, Atlanta, USA, pp. 2235–2245, 2005.

[HO 03] HO T., KOETTER R., MÉDARD M., KARGER D.R., EFFROS M., "The benefits of coding over routing in a randomized setting", *IEEE International Symposium on Information Theory*, Yokohama, Japan, 2003.

[KAT 06] KATTI S., RAHUL H., HU W., KATABI D., MÉDARD M., CROWCROFT J., "XORs in the air: practical wireless network coding", *ACM Conference of the Special Interest Group on Data Communication (SIGCOMM'06)*, New York, USA, pp. 243–254, 2006.

[KOE 03] KOETTER R., MÉDARD M., "An algebraic approach to network coding", *IEEE/ACM Transactions on Networking*, vol. 11, no. 5, pp. 782–795, 2003.

[LE 10] LE J., LUI J.C.S., CHIU D.-M., "DCAR: Distributed Coding-Aware Routing in Wireless Networks", *IEEE Transactions on Mobile Computing*, vol. 9, no. 4, pp. 596–608, 2010.

[LI 03] LI S.Y.R., YEUNG R.W., CAI N., "Linear network coding", *IEEE Transactions on Information Theory*, vol. 49, no. 2, pp. 371–381, 2003.

[NAR 08] NARMAWALA Z., SRIVASTAVA S., "Survey of Applications of Network Coding in Wired and Wireless Networks", *National Conference on Communications (NCC'08)*, Mumbai, India, pp. 153–157, 2008.

[NI 06] NI B., SANTHAPURI N., ZHONG Z., NELAKUDITI S., "Routing with Opportunistically Coded Exchanges in Wireless Mesh Networks", *IEEE Workshop on Wireless Mesh Networks (WiMesh)*, Reston, USA, pp. 157–159, 2006.

[NET 00] Network Coding Bibliography, 2000, http://www.ifp. illinois.edu/~koetter/NWC/index.html.

[OMI 08] OMIWADE S., ZHENG R., HUA C., "Practical localized network coding in wireless mesh networks", *IEEE Conference on Sensor, Mesh and Ad Hoc Communications and Networks (SECON'08)*, San Francisco, USA, pp. 332–340, 2008.

[SIL 07] SILVA D., KSCHISCHANG F.R., "Rank-Metric Codes for Priority Encoding transmission with network coding", *Canadian Workshop On Information Theory (CWIT'07)*, Edmonton, Canada, pp. 81–84, 2007.

[SUN 09a] SUNDARARAJAN J.K., SHAH D., MÉDARD M., MITZENMACHER M., BARROS J., "Network coding meets TCP", *IEEE International Conference on Computer Communications (INFOCOM'09)*, Rio de Janeiro, Brazil, pp. 280–288, 2009.

[SUN 09b] SUNDARARAJAN J.K., DEVAVRAT SHAH M.M., SADEGHI P., Feedback-based online Network Coding, Patent Publication Number: US 2010/0046371 A1, 2009.

[YEU 06] YEUNG R.W., LI S.Y.R., CAI N., ZHANG Z., *Network Coding Theory. Part I: Single Source*, vol. 2, Now Publishers, Hanover, USA, June 2006.

Chapter 2

Fountain Codes and Network Coding for WSNs

2.1. Introduction

Current communication networks have the ability to interconnect billions of devices (computers, phones, PDA, sensors, etc.) and thus enable us to share digital resources independently from their physical location. This network growth has been largely favored by the development of wireless transmission technologies (cellular networks, WLAN, WPAN, WiMax, etc.), which have enabled us to increase the deployment of a large number of communicating terminals at lower cost.

Currently, it is possible to embed on a sensor a module enabling RF transmission of data measured in the environment. In this context, it becomes possible for network sensors to create an intelligent network that is

Chapter written by Anya APAVATJRUT, Claire GOURSAUD, Katia JAFFRÈS-RUNSER and Jean-Marie GORCE.

able to measure and interact with the environment where the aforesaid network is deployed. Such network is called a wireless sensor network (WSN). This chapter aims to show how it is possible to leverage fountain codes and network coding in the context of a WSN.

These sensors have become major elements in every system where the knowledge of the outside environment is required to evaluate the situation and act [PUJ 04]. Each node mainly comprises one or several radio emitters and receivers, one processing unit, and the other power supply. These nodes aim to detect, control, and collect the information of a given geographical area and transfer this information to one or several central entities commonly referred to as *sinks*.

WSNs differ from other wireless communication networks mainly because of two features. The first difference is *energy stress*. In fact, the sensors are often supplied with energy by batteries. Therefore, it is necessary to restrict the energy consumption of the system to increase its lifetime. An important part of the consumption is due to the emission and reception of packets. Thus, it is important to restrict the number and power of transmissions in the whole network to preserve the global resources.

The second difference is linked to *transmission reliability* in this type of network. Because of the deployment mode of this type of network, transmissions are particularly error prone. In fact, these networks can be deployed in harsh environments: poorly known and poorly mastered, difficult to access, in weather conditions that are sometimes extreme (heat, humidity, etc.). Thus, it becomes difficult to predict the type of transmission channel. The sensors themselves can also be prone to system malfunctions.

In this chapter, we show that it is possible to use fountain codes and network coding to overcome these two issues.

In fact, fountain codes are *rateless* codes, naming with no predefined rate. They adapt themselves automatically to any channel by sending an infinite flow of coded packets from the source. Thus, a single end-to-end acknowledgment is sent by the destination once it has decoded all the source packets. Furthermore, it is possible to improve the performances of the transmission (in terms of energy and time) if network coding is carried out by the different relays in the network. This performance gain is obtained by the diversity introduced by network coding at the level of the encoded packets.

This chapter can be divided as follows. Section 2.2 introduces the different families of fountain codes and their main features. Section 2.3 presents the performances of fountain coding in WSN. The impact and performances of network coding associated with fountain coding are detailed in section 2.4. Finally, section 2.5 concludes the chapter.

2.2. Fountain codes

During any transmission of information in a network, the signal undergoes varying levels of perturbation in the transmission channel: path loss, interference, and noise. As a consequence, the received signal is degraded and can lead to errors in the estimation of emitted data.

There exist several ways to enhance reliability in the transmission of data in a network, to overcome imperfections of the transmission channel. This can be carried out as follows:

– At the application layer, by carrying out an interpolation between the previous and the following data. However, it is possible only for applications based on a continuous phenomenon (e.g. for voice).

– At the routing layer, like for instance with protocols such as robust routing [FEI 10], gradient routing [LUK 09], diffusion routing [YE 05], and even virtual coordinate routing [THO 08].

– At the MAC layer with the design of protocols leading to the retransmission of erroneous or lost data: with the help of an access control protocol CSMA/CA [IEE], by using a diagram of standby/awakening [YE 03, POL 04], by using acknowledgments (ARQ/HARQ [COM 84]), by defining retransmission requests through adaptive algorithms [FLO 97, PAU 97, OBR 98], and so on.

– At the physical layer. In this case, we use codes enabling the detection and/or correction of transmission errors. These codes are referred to as forward error correction. They carry out the function of *channel coding*.

The last solution is efficient only if the code type used is adapted to the transmission channel. Its use, thus, requires *a priori* knowledge of the channel. Conversely, the enhancement of reliability of data by MAC layer is more flexible, but generates more traffic to control data reception and retransmission.

In this chapter, we are interested in fountain codes, which have the feature of generating a data flow adapted to every type of transmission channel. These codes thus enable us to decrease the control messages for MAC layer, while preserving its flexibility.

In the following section, we are going to recall a few general concepts on coded data transmission, before detailing fountain codes.

2.2.1. *Generalities*

In the general case, the recipient receives altered data originating from transmission errors. In the case of a binary erasure channel (BEC), data are considered either as perfectly received or lost. In practice, data are lost when:

– either the signal level is too weak to be perceived;

– or the receiver identifies that the received packet is corrupted and erases it.

As a consequence, for a BEC, the coding used must enable the reconstruction of lost data. The coding can be applied at the level of each data symbol to detect, or even to correct, the erroneous symbols. It can also be applied at the level of packets to correct the erasure of some of them and protect the flow of packets.

Thus, in a regular transmission, the emitter breaks the data to be distributed into packets. Then, an identifier is appended to each packet to enable packet reordering at the destination. The packets are then transmitted to the recipient.

Coding consists of generating excess data to introduce redundancy. We define the *rate* of a code as ratio K/N, with K being the initial number of packets (or symbols), and N being the number of transmitted packets (or symbols).

The most striking result on channel coding has been presented by Claude E. Shannon [SHA 48]. He showed that, for any channel, it is possible to transmit the information with an error rate arbitrarily close to zero, as long as the coding rate is smaller than a threshold named *channel capacity*. Unfortunately, the proof he showed did not provide any code that would allow us to reach this limit. Several works have hence been led to find the codes enabling us to get closer to this limit.

The codes conventionally used are

– RS : *reed Solomon* (DVB, ADSL/ADSL2/ADSL2plus);

– Turbo codes (GSM);

– LDPC : *low-density parity check* (DVB-S2 10GBase-T Ethernet, Wi-Fi 802.11n).

However, these codes have three main limitations:

– In the case of multipath diffusion, the encoding rate is set by the worst channel, which penalizes the best located receivers.

– For a fixed code, if the loss of packets is greater than $N - K$, then the received piece of information cannot be used and the transmission must restart from scratch.

– The channel must be perfectly estimated and known by the emitter so that the code output is adapted. This is a problem for channels with quick variations, or for opportunistic networks, because the piece of information is transmitted without a predefined path.

The use of fountain codes allows us to solve these problems. In fact, these codes have the feature of adapting to any channel type: the coded data are generated until the acknowledgment message reaches the emitter. This acknowledgment is emitted by the receiver once it has managed to recover all the information from the packets it has received.

As a consequence:

– In the case of multipath diffusion, the receivers receive and decode the piece of information at their own rhythm, and can be put in standby once they have received everything.

– The quantity of lost packets is no longer critical for decoding, only the number of packets received matters (regardless of their arrival order).

– The channel does not have to be estimated *a priori*.

Hence, the fountain codes enable the reliable enhancement of data transmission in a context where the state of the network is not known, as is often the case in a WSN.

2.2.2. *Families of fountain codes*

Irrespective of the family of code used, before carrying out the coding, the data of the message are decoupled into K fragments. These fragments are then added by XOR in a random way, and integrated into the packet as useful data. The number of fragments present in a coded packet is referred to as a *degree*. To be able to decode the piece of information at the receiver end, we add to each packet a header that contains any of the following:

– the list of fragments present in the packet;

– a binary vector whose each value is at '1' if the corresponding fragment is in the packet, '0' otherwise;

– a seed allowing the receiver to redo the same random draw as the emitter.

The families of fountain codes will differ by their coding and/or decoding algorithms.

2.2.2.1. *Random fountain codes*

The first family of fountain code is referred to as random linear fountain (RLF). It has an optimal rate, i.e. it enables a successful decoding for a minimum number of received packets.

2.2.2.1.1. Encoding of RLF code

Each RLF-encoded packet is generated from the XOR linear combination of fragments uniformly chosen with a probability of $1/2$.

This coding method can be represented by generating matrix $\mathbf{G}$, each column of which represents the fragments contained in a packet. The encoded packets take the form

$$P = \mathbf{G}^{\mathrm{T}} \cdot F \qquad\qquad [2.1]$$

with F being the vector of initial fragments.

For instance, for $K = 3$, we can have the following coding matrix $\mathbf{G}_{K \times N}$ that enables us to create the N first coded packets:

$$\mathbf{G} = \begin{pmatrix} 1 & 0 & 1 & 1 & 0 \\ 1 & 1 & 0 & 0 & 0 \\ 0 & 0 & 1 & 0 & 1 \end{pmatrix}$$

Thus, the first built packet contains XOR between fragments f_1 and f_2.

2.2.2.1.2. Decoding of RLF code

Decoding of RLF code uses the decoding technique of maximum likelihood (ML). If we consider a noiseless case, it amounts to the matrix inversion, with G' as an invertible submatrix of G^T:

$$F = \mathbf{G}'^{-1} \cdot P \qquad\qquad [2.2]$$

According to [MAC 05], to recover the K fragments of information with a success probability of $1 - p_e$, $K + \log_2(1/p_e)$ packets must be received on an average. Thus, this code enables us to asymptotically reach the capacity of the channel, because of a minimum number of transmissions. This is obtained at a price of high decoding complexity (in the order of $\mathcal{O}(K^3)$). Thus, this type of code is adapted to applications that have a strong constraint on the use of the bandwidth, and a weak constraint on the calculation power of the receivers.

2.2.2.2. Luby Transform (LT)

Luby suggested in [LUB 01] another family of code called LT. The objective of this family is to enable a less complex decoding than that of RLF code. Thus, we first present the decoding algorithm followed by the coding algorithm.

2.2.2.2.1. Decoding of LT code

Decoding by belief propagation (BP) is based on the fact that the packets of degree 1 (i.e. containing a single fragment) can be considered as decoded. Thus, with the help of fragments that are already decoded, the decoder decreases at each iteration the degree of coded packets until all the fragments are decoded. The quantity of packets necessary to be able to decode without error is in the order of $K + \epsilon$ with ϵ being the rate of redundancy of the code.

The detailed analysis of the decoding process by BP has been developed in [HAJ 06, PAK 07]. These studies enable us to develop the analytical expression of the error probability of the decoding [KAR 04, MAN 06, MAA 09]. The redundancy rate introduced by LT code was quantified for symmetrical binary and erasure channels [ETE 04, KHA 02].

To better understand the decoding process of LT code, we present a simple example in which the source has at its disposal a message $\mathcal{M}$ to be transmited to the recipient. This message is cut to create $K = 3$ fragments $F = \{f_1, f_2, f_3\}$ of the same size (by adding stuffing bits, if necessary). The source encodes these fragments with the algorithm described in the following section. Encoded packets $P = \{p_1, p_2, ..., p_N\}$ are then transmitted to the recipient. These LT encoding steps are represented in Figure 2.1.

After reception, the recipient starts the iterative decoding process according to the following steps:

  – At the beginning, the decoder receives only packet $p_1 = f_1 \oplus f_2$. Because of the lack of packets of degree 1, the decoding cannot be carried on. The receiver must wait for the reception of additional packets.

  – The second packet $p_2 = f_1$ being of degree 1, we can directly recover fragment f_1. The latter is then erased from the other combinations of available encoded packets, i.e. from packet p_1 in our example. This enables us to decrease the degree of packet p_1 to 1, and hence recover fragment f_2. Then, not having any more fragments of degree 1, the receiver must wait for the arrival of a new packet.

– The decoder then receives packet $p_3 = f_2$ containing the already encoded fragment, this packet does not bring any new information.

– Finally, the fourth packet $p_4 = f_2 \oplus f_3$ arrives at the decoder. Given that we already know f_1 and f_2, this packet allows us to decode the last missing fragment f_3.

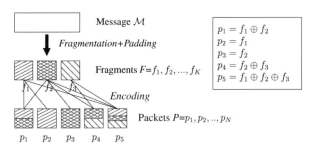

Figure 2.1. *LT encoding*

Figure 2.2. *State of the decoder after the reception of a packet*

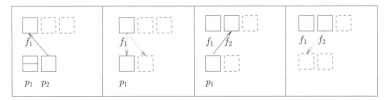

Figure 2.3. *State of the decoder after the reception of two packets*

Figure 2.4. *State of the decoder after the reception of three packets*

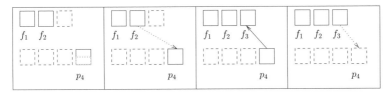

Figure 2.5. *State of the decoder after the reception of four packets*

In this small decoding example, we notice that the decoder requires four packets to be able to decode the three initial packets. We have $N = 4$, $K = 3$. As a consequence, the code rate $R = 3/4$.

2.2.2.2.2. Encoding of LT code

To obtain the best performance with the BP algorithm, the packets must be encoded in a specific way. In fact, we must find a good balance between low-degree packets (to hope to be able to provide enough packets of degree 1 during the decoding algorithm), and high-degree packets (to hope to be able to warrant the presence of all the fragments at the receiver). Theoretically, the ideal soliton distribution (Figure 2.6(a)) suggested by Luby enables us to check these properties.

$$\rho(i) = \begin{cases} \dfrac{1}{K} & \text{for} \quad i = 1 \\[2mm] \dfrac{1}{i(i-1)} & \text{for} \quad 2 \leqslant i \leqslant K \end{cases} \qquad [2.3]$$

Yet, this distribution is not very efficient in practice. The decoding occasionally stops because of a lack of low-degree packets. Furthermore, some fragments do not often appear in the packets. To overcome these problems, Luby suggested the robust soliton distribution (RSD) so that BP decoding efficiently carries on. This is optimal in terms of capacity (Figure 2.6(b)).

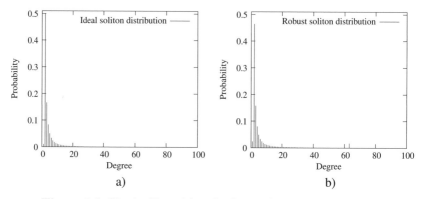

Figure 2.6. *Ideal soliton (a) and robust soliton (b) distribution*

For this, we impose the expected value of the number of packets of degree 1 at each decoding step to be greater than 1:

$$S \equiv c \ln \left(\frac{K}{\delta} \right) \sqrt{K}$$

[2.4]

with c and δ being real numbers. Accordingly, the RSD can be defined by

$$\mu(i) = \frac{\rho(i) + \tau(i)}{Z},$$

[2.5]

with

$$Z = \sum_i \rho(i) + \tau(i),$$

[2.6]

and

$$\tau(i) = \begin{cases} S/ik & \text{for} \quad 1 \leqslant i \leqslant \dfrac{K}{S} - 1 \\[2mm] \dfrac{S \ln(\frac{S}{\delta})}{K} & \text{for} \quad i = \dfrac{K}{S} \\[2mm] 0 & \text{for} \quad i > \dfrac{K}{S} \end{cases} \qquad [2.7]$$

The examples of ideal soliton distribution and RSD are presented in Figures 2.6(a) and (b), respectively. We particularly notice a specific value referred to as a *spike* around the value of $K \approx 60$. These distributions are presented for a code of dimension $K = 100$.

The general reconstruction of K fragments can be done from any combination of $K + \mathcal{O}(\sqrt{K} \log^2(K))$ packets by using an iterative decoder that has a cost in the order of $\mathcal{O}(K \log(K))$ [BEI 07].

The number of symbols required by the decoding tends toward K when dimension K of the code tends toward infinity. In other words, LT code is asymptotically perfect. However, for short code lengths, the redundancy introduced by the code cannot be disregarded with respect to dimension K. Several techniques to optimize the degree distribution are suggested for small K in [HYY 07, LU 09]. We can also consider the dynamic degree adaptation with the help of the feedback channel [HAG 09].

The encoding of LT codes is thus carried out according to Algorithm 1. The distribution of degrees is defined by RSD distribution. The combination of fragments is done in $\mathbb{F}_2^K$ by the addition modulo 2 (XOR). This algorithm stops when the decoder receives enough packets to be decoded or when a maximum timeout is reached.

Algorithm 1: Encoding of LT code

---

while we have not received the instruction to stop **do**
    choose d with the help of RSD distribution
    randomly and uniformly choose d fragments amongst K
    XOR the selected fragments
    send the data obtained with the header
end while

The algorithm being mainly based on random draws, LT code requires a low complexity calculation at the emission ($\mathcal{O}(\log(K))$ operations by packet sent).

2.2.2.3. *Raptor code*

Raptor code derives from LT code, and consists of improving the resistance to noise of LT code (outer code) by precoding (inner code) [SHO 06]. This precoding is based on a linear block code with high rate. The LT code used aims to maintain the weak BP decoding complexity. The precode enables us to assist the decoding of information when the number of packets received is less than K. The choice of precoding differs according to the applications. We find, for instance, the precoding by LDPC codes or Hamming codes.

Raptor code also enables us to improve some features missing in LT codes. Thus, they enable better performances for a small dimension K. On the other hand, it can be adapted in a systematic version [FRE 07, SEJ 09]. In this case, the first K fragments are sent without being encoded, then the encoded redundancies are sent. This enables us to decrease the transmission latency, in particular when the erasure on the channel is weak because the decoder can start to recover the fragments sent without having to wait for the end of the transmission.

The coding and decoding complexities of Raptor code are linear in time in the order of $\mathcal{O}(\log(K))$ operations by packets.

2.2.2.4. *Code complexity*

The comparison of different fountain codes is done in Table 2.1 [TIR 09], where N represents the mean number of packets required by the decoder to be able to recover the K original fragments of information.

2.3. Fountain codes in WSNs

The perturbations in a wireless channel lead to the degradation of the quantity of transmissions. The coding solutions are interesting candidates to enhance the reliability of the transmission for such transmission channels. The interest in fountain codes for a WSN can be mainly summarized by the following two points:

– They considerably decrease the number of acknowledgments.

– They intrinsically adapt to losses in the network.

Thus, they enable the decrease in energy consumption of the network and increase its lifetime.

Code	Encoding	Decoding	N
Reed-Solomon	$\mathcal{O}(K)$	$\mathcal{O}(K^3)$	K
LDPC	$\mathcal{O}(1)$	$\mathcal{O}(K)$	$K + \epsilon$
RLF	$\mathcal{O}(K)$	$\mathcal{O}(K^3)$	$K + \mathcal{O}(1)$
LT	$\mathcal{O}(\log(K))$	$\mathcal{O}(K\log(K))$	$K + \mathcal{O}(\sqrt{K}\log^2(K))$
Raptor	$\mathcal{O}(1)$	$\mathcal{O}(K)$	$K + \epsilon$

Table 2.1. *Encoding and decoding cost for different types of code*

In WSNs, fountain codes in particular have been used in the field of distributed storage to reliably diffuse the data to all the sensors in the network. We can cite several examples in this context that stem from RLF codes [LIN 07a], LT codes

[LIN 07b], growth codes (a variant of LT code) [KAM 06], Raptor codes [ALY 08], and so on. The aim of the distributed storage is to distribute K fragments of information among the sensors by disseminating them so that any node can recover the complete information after having received $K + \epsilon$ encoded packets from its neighbors. Outside data storage, they are also used to reprogram sensors [ROS 08].

In this chapter, we present the main advantages of fountain codes for an end-to-end transmission in a WSN. The optimal performances being obtained when the dimension of the code is large, fountain codes are particularly adapted to network applications that require the transfer of a large quantity of data. In the case of WSNs, the data sent are often small. In this case, the redundancy of the code cannot be disregarded with respect to its dimension. In fact, the use of such codes leads to an additional overload due to the:

– widening of the packet header, and

– redundancy rate of the code.

In the following section, we study the compromise between the overload introduced by the codes and the benefit obtained in terms of reliability enhancement of the transmissions.

2.3.1. *Implementation*

The acknowledgment message is sent by the recipient at the time of decoding of K fragments. It is an end-to-end acknowledgement between the application and the routing layers of the protocol stack as shown in Figure 2.7. When a fountain code is used, it is not necessary to acknowledge each packet hop-by-hop at the level of MAC layer in the case of a multihop transmission.

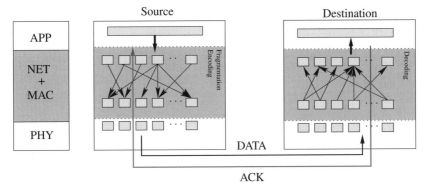

Figure 2.7. *Implementation of the coding layer in the protocol stack of the sensor*

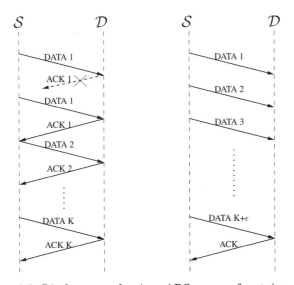

Figure 2.8. *Discharge mechanism: ARQs versus fountain codes*

2.3.2. *Protocol of reliability enhancement: ARQs versus fountain codes*

To check if the use of small dimension codes is really beneficial in a sensor network, we compare an end-to-end transmission using fountain codes and a conventional transmission using a protocol solution of the ARQ type. Figure 2.8 shows that with ARQ solution, a retransmission is done for each data packet lost, whereas with a fountain code, an acknowledgment is only sent once all the fragments are decoded.

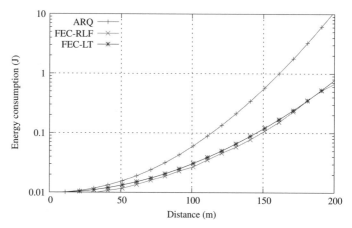

Figure 2.9. *The energy consumption for point-to-point transmission with $K = 100$*

Here we are interested in the energy consumption of the direct transmission between a source and its recipient on a wireless channel. As stated earlier, the energy consumed is a vital criterion for a WSN. The evolution of this consumption to enhance the reliability of the network with either an ARQ protocol or an RLF code and an LT code is presented in Figure 2.9 (see [APA 09] for more details). We obviously notice that the energy consumed rapidly increases with the distance between the source and the recipient. The most interesting thing is that the energy required to enhance the reliability of the transmission with a fountain code is much smaller than that for ARQ. RLF code gives a performance slightly higher than LT code because of its reduced redundancy. The advantage of the fountain code is easily explained by the decrease in the number of acknowledgments and retransmissions when the channel condition for the ACK is bad. Thus, the process of acknowledgment becomes less sensitive to losses linked to the quality of the feedback channel. We can conclude here that even for a small dimension code (here $K = 100$), the fountain code enables us to improve the energy performances of a wireless transmission.

2.3.3. *Discharge and overflow*

Owing to the working mode of a transmission by fountain code, it becomes relevant for us to ask the question of the stopping time of the fountain for a multihop network. In fact, the source permanently transmits packets until reception of the acknowledgment. However, for a multihop communication, the transmission time of the ACK from the recipient to the source cannot be disregarded. It is the time during which the source and the relays keep emitting coded packets. Clearly, knowing that the fountain codes have initially been modeled for a point-to-point link, scaling up these codes to a network is not systematic.

To model the transmission along a route in a sensor network, we are interested here in a linear multihop network. In this network, the acknowledgment cannot arrive instantaneously at S. Thus, during the time the acknowledgment is being sent, the source keeps emitting new coded packets, which are themselves retransmitted by the intermediate relays. These packets are *de facto* totally useless to decoding. We refer to this emission overload as *overflow*. This overflow mainly depends on the following parameters:

  – network length n_h expressed in number of hops;

  – mean progression h of acknowledgment packets;

  – relaying strategy of intermediate nodes.

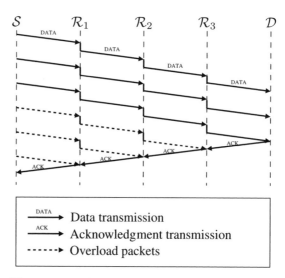

Figure 2.10. *A transmission sequence for $n_h = 4$ and a progression $h = 1$*

Progression h represents the fact that an ACK packet can progress h times more quickly than a data packet. This is the case for some MAC layers that favor the relaying of an acknowledgment by putting it at the top of the queue (preemption of ACK packets over DATA packets). We suggest enforcing an overflow criterion for the following three relaying strategies:

– block relaying (CB): each relay R_i waits to decode all the K fragments received before transmitting a new flow of re-encoded messages that are made of K fragments toward R_{i+1}. The acknowledgment is sent from hop to hop by each relay (from R_i to R_{i-1}).

– packet relaying (CP): each relay R_i relays a coded packet once it is received; only the recipient decodes the packet. The acknowledgment is here sent end to end when the message is decoded.

– hybrid relay (CH): is a hybrid combination of the previous two strategies. In this case, each relay R_i starts relaying the received message toward R_{i+1} and attempts, in parallel, to decode the K fragments. As soon as it has decoded them, it sends an acknowledgment to R_{i-1}.

From these three modes of acknowledgment propagation, it is possible to model the overflow analytically. For the case where the ACK packet is acknowledged from hop to hop, like in the block-relaying scenario, the overflow is minimum and is in the order of $\mathcal{O}(n)$ packets. For the case where the acknowledgment is done end to end like in the case of packet relaying, the acknowledgment is generated only by the recipient and in this case, the overflow is in the order of $\mathcal{O}(\frac{n_h^2}{h})$ packets [APA 11c].

To focus only on the overflow effects, we present only the results relative to the RLF code, not to account for the transmission overhead linked to the redundancy of the LT code. We consider a configuration where the ACK packets are processed with absolute priority in the queue of a node. The performance of the transmission is measured by counting the number of transmissions done by the source and by gathering them into effective transmissions (DATA packets plus one ACK), DATA transmissions in overflow, and useless ACK transmissions. These measures are carried out for $n_h \in \{5, 10, 15, 20\}$.

According to Figure 2.11, we can observe that the number of transmissions increases with the number of hops. Widening the network leads to a quick increase in the overflow of DATA packets. The minimum overflow is obtained for a hop-by-hop acknowledgment as predicted. However, this approach implies that all the relays are able to decode, which is not necessarily the case in a sensor network. For $n_h = 20$, passive transmission is no longer possible due to the important overflow measured. The hybrid relay enables us to obtain an interesting compromise between the two CB and CP scenarios since it enables us to decrease the overflow when the number of hops increases, without requiring all the sensors to decode messages.

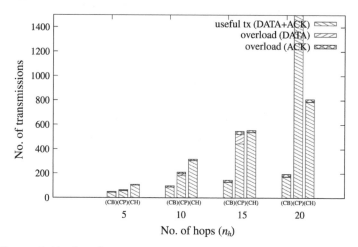

Figure 2.11. *Overflow measured for the three relaying strategies for $n_h = 5, 10, 15, 20$ and $K = 5$*

We can conclude that for delay-tolerant networks, hop-by-hop decoding and encoding is preferable because this strategy avoids useless transmissions. However, for energy- and delay-constrained networks, the choices to be made are not obvious. If the number of hops is smaller than about $n_h \leq 15$, passive relaying is preferable as it does not require any decoding. In addition, we can call for a hybrid strategy CH by imposing that some nodes must completely decode and encode the information before transmitting it to the following node. This enables us to avoid the accumulation of a number of useless transmissions through a large number of hops.

2.4. Fountain codes and network code for sensor networks

As we have seen in the previous section, fountain codes mainly enable us to decrease the energy consumption of a multihop communication by their end-to-end acknowledgment mode. For a multihop communication, if the links are strongly degraded (i.e. undergo a high packet error rate), the transmission of an encoded flow with a fountain code suffers from an important increase in the delay and in the end-to-end energy consumed. In fact, the source must send more packets to compensate for the losses, packets that are relayed as far as possible by the network. As shown in the following section, network coding enables us to improve the transmission performances of an LT flow. In fact, if each relay recombines coded packets between them, it is going to diversify the information transmitted during the progression of the packets in the network. More specifically, if some fragments have been the victims of an erasure in the network, it is possible to increase their arrival chance if a relay has registered them in its memory and recombines them later with other packets. This enables us to understand the extent to which network coding enables us to decrease the transmission time and the energy consumed by an LT flow. However, as we detail it in this section, network coding must be applied with care to obtain such performance gains.

We are interested in a network coding where the coded packets can be combined by a simple XOR operation carried out by a relay. The main question is the choice of code *per se*. The chosen code can be represented by an algorithm of packet combination present in the memory of the node. This algorithm aims to define how many memory packets are combined at the time of this operation (at the reception of a packet, at regular time intervals, etc.). It also determines if the packets are conserved in the memory or not following the sending. We see that these parameters modify the end-to-end performances of the transmission, either by improving them or by penalizing them[1].

Section 2.4.1 starts with a presentation of the impact of a very easy network coding algorithm on the quality of transmissions. We update a degradation of performances due to the impact of network coding on the distribution of the degree of packets of LT flow to the destination. Section 2.4.2 suggests different algorithms with the aim of taking advantage of the benefits of network coding while preserving the distribution of the degree. The latter are illustrated in section 2.4.3 for a linear multihop network and a sensor network.

2.4.1. *Impact of network coding on the degree distribution of an LT flow*

2.4.1.1. *XOR network coding and LT code*

In our analysis, we consider only the impact of network coding on an LT code. This choice is motivated by the weak decoding complexity of LT code compared with RL code, which makes it interesting for its implementation in a WSN.

We consider a wireless multihop sensor network where the source emits an LT flow of packets. Each encoded packet $p \in \mathbb{F}_2^N$ of size N is of the form $p = \{t||m\}$ where $t \in \mathbb{F}_2^K$ is the header of K bits describing the linear combination in $\mathbb{F}_2$, and $m \in \mathbb{F}_2^M$ represents the data of the packet of size M. The nodes of the network apply a network code where the coefficients are chosen in the finite field $\mathbb{F}_2$. For this, each node of the network can, if necessary, register packets in a proper memory. A new

---

1 To obtain more details on the results presented in this section, we suggest the reader to consult references [APA 11a] and [APA 11b].

packet is generated by combining the packets of the memory with the help of XOR operator.

$$p_{xor} = \oplus_{i \in \mathcal{B}} \ \alpha_i \cdot p_i \quad \text{with} \quad \alpha_i \in \{0, 1\} \qquad [2.8]$$

where $\mathcal{B}$ represents all the packets present in the memory of the node. In what follows, we are interested in the particular case where we combine only two packets of the memory between them. In this case, the emitted packet is the binary sum of two packets $p_A \in \mathcal{B}$, and $p_B \in \mathcal{B}$ as illustrated by Figure 2.12.

$$p_{xor} = p_A \oplus p_B = \{t_A \oplus t_B \| m_A \oplus m_B\} \qquad [2.9]$$

We choose to represent the header by K binary values signifying the presence or absence of fragments in the coded packet. This representation mode of the header enables us to directly carry out the same XOR combination on the header and the data, without previous conversion.

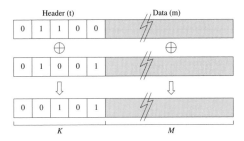

Figure 2.12. *XOR combination of two LT packets*

To have the fastest decoding of the encoded packets by an LT code, all the packets received by the destination must present a degree distribution that follows the RSD (see Figure 2.13(a)). If this distribution is not respected, the source must transmit more packets so that the decoding is possible by the growth propagation algorithm. It hence ensues therefrom a lengthening of the transmission time and an increase in the energy consumed by the network.

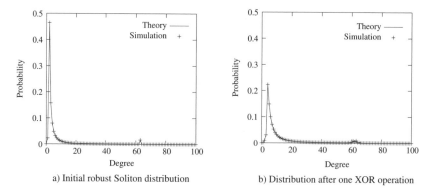

Figure 2.13. *Distribution of the degrees after XOR combination of two packets for* $K = 100$

In what follows, we denote by d_A, d_B, and d_{xor} the degree of packets p_A, p_B, and p_{xor}, respectively. We recall that the degree is given by Hamming's weight of the header according to $d = W_H(t) = \sum_{i=1}^{K} t_i$. If we take a closer look at the XOR combination effect on the degree of a resulting packet ($d_{xor} = W_H(t_A \oplus t_B)$), we notice that the latter is modified in the following way:

– If the two packets p_A and p_B do not have any fragment in common, the resulting degree is the simple sum of the degrees of p_A and p_B: $d_{xor} = d_A + d_B$.

– If the two packets possess o fragments in common, there is *collision* for all these fragments, which leads to the deletion of these fragments in the resulting packet. The resulting degree is hence expressed simply as $d_{xor} = d_A + d_B - 2o$.

From this observation, it is possible to infer the degree distribution resulting from two XOR combined LT packets. If

$$p_o(o|d_A, d_B, K) = \begin{cases} \dfrac{C_{d_A}^o C_{K-d_A}^{d_B - o}}{C_K^{d_B}} & \text{for } o \leq \min(d_A, d_B) \\ 0 & \text{otherwise} \end{cases} \quad [2.10]$$

is the probability that there are exactly o collisions after the combination of p_A and p_B, we can infer the probability to obtain degree d_{xor} knowing d_A, d_B, and o:

$$p_{dxor}(d_{xor}|d_A, d_B, o) = \begin{cases} p_o(o = \frac{d_A + d_B - d_{xor}}{2}|d_A, d_B, K) & \text{if } o \in \mathbb{N} \\ 0 & \text{otherwise} \end{cases}$$

[2.11]

By applying the total probability law on all the degrees d_1 and d_2, equation [2.11] becomes

$$p_{dxor}(d_{xor}) =$$

$$\sum_{d_A=1}^{K} \sum_{d_B=1}^{K} p_{dA}(d_A)p_{dB}(d_B)p_o\left(o = \frac{d_A + d_B - d_{xor}}{2}|d_A, d_B, K\right)$$

[2.12]

where $p_{dA}(d_A)$ and $p_{dB}(d_B)$ correspond to the probability of obtaining degrees d_1 and d_2 for the initial packets.

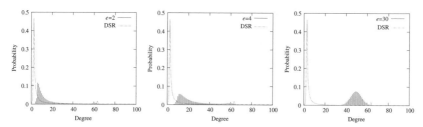

Figure 2.14. *Evolution of the degree distribution after e successive XOR combinations for $K = 100$*

The distribution of the degree of p_{xor} is illustrated and compared to the RSD in Figure 2.13. We clearly see that this simple network coding increases the mean value of the degree of the packets. This result can be extended to the case where e successive combinations of $e+1$ LT packets are carried out. For the sake of brevity, we omit the analytical derivation of the expression and present only the graphical results in Figure 2.14 for values of $e = \{2, 4, 30\}$. The expected value of the degrees tends toward $K/2$ when e increases, as shown by Figure 2.15. At the limit, the degree distribution tends toward the distribution of a linear random RLF code.

2.4.2. *Design a network code for LT code*

The drift of the packet degree distribution due to network coding largely modifies the transmission performances. In fact, the RSD no longer being fulfilled, the BP decoding algorithm does not receive enough packets of small degree to decode all the fragments emitted at the source. A first solution is to use a linear random code for which the degree distribution remains unchanged by network coding. However, this solution requires ML decoding of a $\mathcal{O}(K^3)$ complexity much greater than the complexity of a BP algorithm in $\mathcal{O}(K \log(K))$.

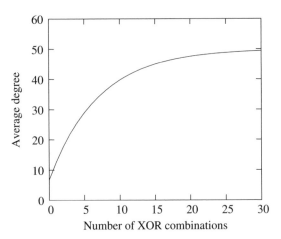

Figure 2.15. *Expected value of the degree after e successive XOR combinations for $K = 100$*

For applications requiring a reduced decoding complexity, it is preferable to work with LT code. In this case, we must overcome the drift of the degree distribution of LT code by successive combination of packets. The solution presented here consists of searching for a network code that maintains the end-to-end degree distribution. The aim of this section is to make the reader aware of the design of such combinational algorithms by presenting two possible algorithmic solutions to this problem. These solutions are first illustrated by a linear multihop network and then extended to a larger-scale sensor network.

2.4.2.1. Solutions of network coding

As previously described, each relay possesses a memory having a set $\mathcal{B}$ of received packets. The relays can combine two packets or more with the help of an XOR operation. This combination takes place at each new reception of packets. Algorithm 2 presents the main actions of a relay at the reception of a packet.

Algorithm 2: Macro-algorithm of network coding at the level of a relay

---
1: **if** Packet reception **then**
2:    Select the set $\mathcal{S}$ of packets of $\mathcal{B}$ so that d_{xor} follows the Robust
   Soliton distribution, with $d_{xor} = W_H(t_{xor}) = W_H(\oplus_{i \in \mathcal{S}} t_i)$
3:    Create $p_{xor} = \oplus_{i \in \mathcal{S}} p_i$
4:    Emit p_{xor}
5:    Erase the combined packets of $\mathcal{B}$
6: **end if**

Instruction 2 of this algorithm lies at the heart of the problem: how to choose the set of packets $\mathcal{S}$ that minimizes the drift of the packet degree distribution at the destination. Hence, we must choose both the number of packets of $\mathcal{B}$ and the most relevant packets to be combined. Two solutions are presented, one from the statistical analysis relative to the degree distribution [NOD 10] and another algorithmic method of construction of a packet of targeted degree d_{obj} [APA 11b].

2.4.2.1.1. Theoretical approach

In this approach, we restrict the size of $\mathcal{S}$ to two packets p_A and p_B chosen in $\mathcal{B}$. The first packet p_A chosen is the last packet received of degree $d_A = 1$ (see [NOD 10] for the explanation of this choice). The question is to determine the degree of p_B to obtain a packet p_{xor} that follows the RSD.

Algorithm 3: Theoretical approach

---

p_A ⇐last packet received in the memory whose degree is equal to d_A

if p_A does not exist **then**

  exit

else

  d_B ⇐degree chosen according to $p_{choice}(d_B)$

  p_B ⇐packet of degree d_B randomly chosen in $\mathcal{B}$

  if p_B does not exist **then**

    exit

  end if

end if

p_{xor} ⇐ p_A XOR p_B

transmit p_{xor}

To select them, we rely on the calculation of probability $p_{dxor}(d_{xor})$ to obtain a packet of degree d_{xor} after XOR combination of two packets (see equation [2.13]). Contrary to the previous calculation, we want that p_{xor} follows the RSD.

Thus, $p_{dxor}(d_{xor})$ is known and given by the RSD. Knowing d_A, we have the following equality:

$$p_{dxor}(d_{xor}|d_A) = \sum_{d_B=1}^{K} p_{dB}(d_B) p_o \left(o = \frac{d_A + d_B - d_{xor}}{2} \bigg| d_A, d_B, K \right)$$

[2.13]

Here, $p_{dB}(d_B)$ is the probability of choosing a packet of degree d_B. Knowing $p_{dxor}(d_{xor}|d_A)$ and d_A, it is possible to determine the probability of choice of a packet of degree d_B by solving an inverse problem. In what follows, we refer to $p_{dB}(d_B)$ as $p_{choice}(d_B)$. Hence, we have K unknowns $p_{choice}(d_B), d_B \in \{1, ..., K\}$, which represent the distribution of the probability of choice of degree d_B enabling us to obtain a packet p_{xor} that follows the degree distribution of the Robust Soliton, knowing d_A. The solving of the inverse problem relies on a matrix inversion such as presented in [NOD 10]. The obtained distribution $p_{choice}(d_B)$ is represented in Figure 2.16.

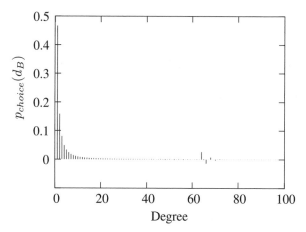

Figure 2.16. *Distribution of the probability of choosing degree p_B knowing $d_A = 1$*

In practice, the relay is going to search for the last received packet of degree 1 in its memory (see Algorithm 3). Then it is going to follow the statistics of p_{choice} for the selection of d_B and then it randomly selects a packet of degree d_B in $\mathcal{B}$. Then it emits the combined version of the two packets before erasing them in $\mathcal{B}$.

2.4.2.1.2. Heuristic approach

This solution does not restrict the number of packets selected in $\mathcal{S}$ and suggests a heuristic approach of selection of packets that enables us to generate a packet p_{xor} whose degree follows the RSD. Algorithm 4 combines the packets at each iteration to get the degree of p_{xor} closer to the target degree d_{obj}. The target degree is obtained from the RSD.

Owing to its structure, it is difficult to create small degrees packets (mainly of degrees 1 and 2) by combining LT packets. Thus, a modified version is suggested in [APA 11b] that applies Algorithm 4 to packets of degrees 1 and 2 only with a given probability ξ. This modification favors the transmission of packets of small degree to be more faithful to the RSD.

2.4.3. *Application to multihop sensor networks*

2.4.3.1. *Multihop linear networks*

In this section, we first consider a simple linear network with $N+2$ nodes: a source S, N intermediate nodes $R_{i,i\in\{1,..,R\}}$, and a recipient D (see Figure 2.17). The route is fixed and is followed hop-by-hop: source S sends to R_1, then R_1 relays to R_2, ..., then R_{N-1} relays to R_N, finally R_N transmits to D. The transmission with two or more hops is not possible: R_i cannot directly receive packets from R_j with $j \leq i - 2$. These nodes are linked by channels that we have assumed neither with erasure nor with interference, and whose probability of success transmission is $\gamma = 1$.

Algorithm 4: Heuristic approach

1: **if** Reception of a packet **then**
2:      $d_{obj} \Leftarrow$ targeted degree according to the Robust Soliton distribution
3:      $p \Leftarrow$ last received packet in the internal memory of degree d_p
4:      $i \Leftarrow 0$
5:      **while** $i <$ maxround **do**
6:          $p_{rand} \Leftarrow$ packet randomly chosen in $\mathcal{B}$
7:          $p_{xor} \Leftarrow p \oplus p_{rand}$
8:          **if** d_{xor} is closer to d_{obj} than d_p **then**
9:              $p \Leftarrow p_{xor}$
10:             **if** $d_{xor} = d_{obj}$ **then**
11:                 exit
12:             **end if**
13:         **end if**
14:         $i \Leftarrow i+1$
15:     **end while**
16:     Transmit p_{xor}
17:     Erase the combined packets in p_{xor}
18: **end if**

Figure 2.17. *Linear multihop network*

Figures 2.18(a) and (b) present the final degree distribution obtained for a linear network with a single relay (N=1) for the theoretical approach and the heuristic approach, respectively. An LT fountain code of dimension $K = 100$ has been used. The results presented hereafter have been obtained for the sending of 100 messages from S to D. The size of the queues is considered to be unlimited in this implementation.

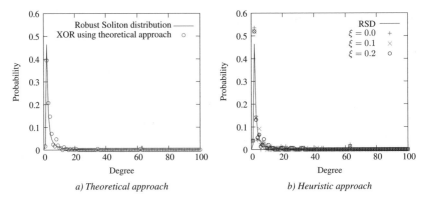

a) Theoretical approach   b) Heuristic approach

Figure 2.18. *Distribution of degrees without and*
with network coding $N = 2$

According to Figure 2.18, we notice that the heuristic approach generates a degree distribution closer to the RSD, mainly for small degrees. This is a consequence of probability ξ of not combining small size packets. In fact, it is difficult to generate small size packets by XOR combination, hence letting through packets of degree 1 or 2 without combining them improves the decoding performance. The mean square error between the results of the heuristic approach and DSR distribution for $\xi = 0.0, 0.1$, and 0.2 is equal to 0.095, 0.089, and 0.082, respectively. It is for a value of $\xi = 0.2$ that the square error is minimum, and it is hence with this value that we apply the heuristic approach in the rest of the chapter. The performance of the theoretical approach is not very high because it relies on the systematic selection of a packet of degree 1, which modifies the degree distribution accordingly.

For a longer network, we are interested in performance indicators such as end-to-end transmission delay (transmission delay but mainly decoding delay) and energy consumed by the network. For this, it is enough for a first estimation to measure the number of packets that the source needs to emit before the destination can successfully decode the message. This metric takes into account both the losses on different links and the quality in terms of degree distribution of the emitted packets. The impact of the number of hops in the networks (i.e. $N + 1$ relays) on the number of messages emitted by the source before decoding is represented in Figure 2.19. The performance of the two approaches is compared to the performance of relaying a simple packet (passive relaying) and to the strategy where all the relays await the decoding of the message before re-encoding it to be transmitted to the following node. Clearly, the heuristic method performs better than the theoretical approach because it enables us to largely improve the transmission performances when the number of hops increases.

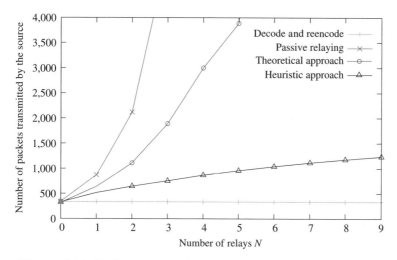

Figure 2.19. *Performances of theoretical and heuristic approaches as a function of the size of the linear network*

2.4.3.2. *Sensor networks*

After having studied the contribution of fountain code and network coding for a linear network, we are now interested in extending this study to the case of a complete two-dimensional network. In this network, we aim to enhance the reliability of a transmission toward the sink by decreasing the energy consumption and the transmission delay. For this, we show that it is possible to take advantage of the spatial diversity of a gradient diffusion and of the diversity introduced by the combination of packets obtained by LT codes and network coding. In this section, we reuse the network coding using the heuristic approach presented earlier.

Gradient broadcasting is an interesting diffusion routing for a sensor network because it requires only a minimum self-organization stage and enables us to increase the robustness of the transmission in difficult environments where the loss rate is high [YE 05, THO 08, JAF 09]. This is a routing that associates a cost Q_i to each node i of the network proportional to a measure of its distance to the sink. All the costs form a decreasing gradient field. The sink lies at the minimum of the field and possesses the minimum cost ($Q_i = 0$). This cost can be obtained from diverse distance metrics such as the number of hops and Euclidean distance to the sink. In our study, we use the Euclidean distance between a relay R_i and recipient D (i.e. the sink). The association of the cost to the node is carried out during an initialization stage: initially, the sink diffuses an ADV message with a current cost Q_c that is equal to 0. Node j which hears this message initializes its cost Q_j to the sum between the current cost Q_c and the distance between its position and the position of the node which has sent this message. Then, it broadcasts this message to other nodes within reach by updating the current cost with its proper cost: $Q_c = Q_j$. This step is repeated until all the nodes are associated with their costs (see [YE 05] for more details).

In the stage of data transmission toward the sink, the packets of a source S are broadcast with the help of previously established costs and follow the direction of decreasing costs to reach their destination. At the beginning of the communication stage, each node broadcasts its packet by associating a cost in the packet header Q_p equal to its cost Q_i. When its neighbors at one-hop away receive these packets, they rebroadcast them if and only if their proper cost is less than the packet cost: $Q_i < Q_p$. At the time of rebroadcast by a sensor k, the packet cost is updated with the sensor cost k: $Q_p = Q_k$. To decrease the multiple retransmission of the same packet, we impose that the sensors do not transmit the same packet emitted by the source twice. This very easy relaying technique generates an important additional transmission cost (redundancy). Different approaches have been suggested to provide a compromise between the robustness introduced by this redundancy and the energy consumption of the network (see [YE 05, JAF 09]).

In this chapter, we suggest introducing the network coding algorithm using the heuristic approach presented earlier in the gradient broadcasting routing. For this, the source emits an LT flow of dimension K. Each node possesses a FIFO-type queue of maximum size $F_{max} = K$. To be able to control the quantity of redundancy generated by this diffusion, we introduce a first decision in the form of a uniform *relaying rate* τ for all the relays of the network. Thus, if the cost of relays k is less than the cost of the received packet Q_p, it will broadcast the packet with a probability τ.

To be able to analyze the impact of network coding, we introduce a second parameter: probability ν of applying network coding. Thus, if k decides according to probability τ of diffusing the message, it will broadcast a combined version of it according to Algorithm 4 with probability ν. Otherwise, it retransmits the packet without coding. The relaying algorithm used by a relay is given by Algorithm 5.

We have hypothesized in this section that the acknowledgment message once sent by the recipient does not undergo any loss. Hence, we have disregarded in this study the additional cost introduced by the acknowledgment mechanism. The diffusion of the ACK message of K packets at the end of the transmission can be done at the same time as the update of the cost field by including it in the ADV message. In the case where the recipient does not receive enough packets that allow him to successfully decode the message, the source continuously transmits the encoded packets as long as there is enough power.

Algorithm 5: Relaying macro-algorithm for a relay R

$F_{max} \Leftarrow K$
if R receives a packet p **then**
    Add p in FIFO of size F_{max};
    if Test on τ == true **then**
        if Test on ν == true **then**
            Apply algorithm 4
        else
            $p_{new} = p$
        end if
    end if
    transmit p_{new}
end if

The simulation results presented hereafter have been obtained with discrete event simulator WSNet [WSN]. The sensor features are defined by the specifications of radio device TI CC1100. The protocols of physical and MAC layers follow standard IEEE802.15.4. For this study, we have hypothesized the presence of orthogonal channels (no interference between sensors) and an AWGN channel model. We consider a network made of 50 sensors uniformly distributed in a 500×500 m space. The mean number of neighbors for each sensor is about three nodes. The simulation results in this section are obtained for the transmission of $M = 50$ messages for a source S toward D. Each message is encoded with an LT code of dimension $K = 100$. We have varied the relaying rate τ between 0 and 1 and probability of network coding ν also between 0 and 1.

We measure the mean success rate for the 50 emitted messages as a function of the relaying rate value in Figure 2.20, for different values of ν. We observe three distinct zones in this figure. The first zone is the unreliable zone where the success rate is very low for $\tau < 0.4$. In fact, in this zone, the relaying rate for the relay is not high enough to ensure the connectivity of the network. For the case where $\tau > 0.6$, the success rate is almost equal to 1. In this case, the relaying rate is adequate to ensure the connectivity of the network. We have observed a transition zone around the value $\tau \approx 0.5$. In this zone, we observe a quick increase in the success rate. In what follows, we refer to this zone as *transient zone* where the network connectivity is sensitive to the coding choice by the relays. Because of the unreliability of the channel and the variability of the connectivity, we have observed that using a systematic network coding ($\nu = 1$) strongly improves the success rate with respect to $\nu = 0$. Network coding, in this context, enables us to improve the reliability of the network with respect to the case without coding.

For a relaying rate $\tau = 0.5$, the improvement of the success transmission rate by network coding is translated into a decrease of 50% of the quantity of energy consumed and a decrease by a factor of 10 of the end-to-end transmission delay. These results are particularly encouraging for sensor networks. In fact, it is particularly interesting for the global consumption of the network to work with a weak relaying rate of the network (or with a smaller density of sensors) while compensating for the loss of connectivity by using network coding.

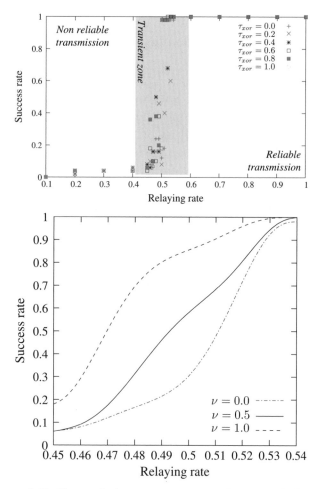

Figure 2.20. *Transmission success rate as a function of relaying rate τ for several values of ν. The bottom figure presents a zoom on the transient zone after smoothing with Bezier curves*

2.5. Conclusion

The first section of this chapter emphasized the interest of fountain codes for enhancing the reliability of transmissions in a multihop sensor network, including when the quantity of information to be transmitted is moderate. From a linear network scenario, we have shown that these codes enable, in particular, the decrease of control messages and a natural adaptation to unpredictable packet losses. However, an overflow phenomenon in the use of such codes appears when we acknowledge only the end-to-end transmissions, and we have suggested and evaluated some solutions to decrease this overflow effect in the network.

In the second section of this chapter, we were interested in the combination of these fountain codes with a network coding, to take advantage spatially of the multiple packet receptions. Network coding presents a major advantage for sensor networks in light of the results presented in the last section of this chapter: it enables us to improve the transmission rate for a network in connectivity limit. When the sensors do not have access to an infinite energy source, network coding becomes a tool that is of particular interest for extending the network lifetime. In fact, at the end of life, such a network sees its nodes die and the connectivity severely decreases. It is there that the packet diversity introduced by network coding enables us to decode the messages transmitted with a weak connectivity. Furthermore, this network coding does not modify the transmission performances when the connectivity is sufficient. It can thus be deployed from the creation of the network without adding additional load to its work. Let us note, however, that these protocols lead to an additional energy cost linked to the reception, since all the potential relay nodes listen simultaneously.

2.6. Bibliography

[ALY 08] ALY S., KONG Z., SOLJANIN E., "Raptor codes based distributed storage algorithms for wireless sensor networks", *IEEE International Symposium on Information Theory*, pp. 2051–2055, July 2008.

[APA 09] APAVATJRUT A., GOURSAUD C., GORCE J.-M., "Impact des Codes Fontaine sur la consommation d'énergie dans les réseaux de capteurs avec prise en compte d'une couche MAC réaliste", *Actes de GRETSI*, Dijon, France, September 2009.

[APA 11a] APAVATJRUT A., De l'usage des codes fontaines dans les réseaux de capteurs multisauts, PhD thesis, INSA-Lyon, Villeurbanne, France, July 2011.

[APA 11b] APAVATJRUT A., GOURSAUD C., JAFFRÈS-RUNSER K., COMANICIU C., GORCE J.-M., "Toward increasing packet diversity for relaying LT fountain codes in wireless sensor networks", *Communications Letters, IEEE*, vol. 15, no. 1, pp. 52–54, January 2011.

[APA 11c] APAVATJRUT A., JAFFRÈS-RUNSER K., KATIA E., GOURSAUD C., LAURADOUX C., "Overflow of fountain codes in multi-hop wireless sensor networks", *Proceedings of the 22nd IEEE International Symposium on Personal, Indoor and Mobile Radio Communications (PIMRC)*, pp. 995–999, September 2011.

[BEI 07] BEIMEL A., DOLEV S., SINGER N., "RT oblivious erasure correcting", *IEEE/ACM Transactions on Networking*, vol. 15, pp. 1321–1332, December 2007.

[COM 84] COMROE R., COSTELLO D. J., "ARQ schemes for data transmission in mobile radio systems", *IEEE Journal on Selected Areas in Communications*, vol. 2, no. 4, pp. 472–481, July 1984.

[ETE 04] ETESAMI O., "Relations between belief propagation on erasure and symmetric channels", *Proceedings of International Symposium on Information Theory (ISIT)*, Chicago, USA, IEEE, p. 209, 2004.

[FEI 10] FEI Y., Reliable and time-constrained communication in wireless sensor networks, PhD thesis, INSA-Lyon, 2010.

[FLO 97] FLOYD S., JACOBSON V., LIU C.-G., MCCANNE S., ZHANG L., "A reliable multicast framework for light-weight sessions and application level framing", *IEEE/ACM Transactions on Networking*, vol. 5, no. 6, pp. 784–803, December 1997.

[FRE 07] FRESIA M., VANDENDORPE L., "Distributed source coding using raptor codes", *Proceedings of IEEE Global Telecommunications Conference (GLOBECOM)*, pp. 1587–1591, November 2007.

[HAG 09] HAGEDORN A., AGARWAL S., STAROBINSKI D., TRACHTENBERG A., "Rateless coding with feedback", *Proceedings IEEE Computer and Communications Societies (INFOCOMM)*, Rio de Janeiro, Brazil, IEEE, pp. 1791–1799, 2009.

[HAJ 06] HAJEK B., "Connections between network coding and stochastic network theory", *Proceedings Stochastic Networks Conference*, June 2006.

[HYY 07] HYYTIÄ E., TIRRONEN T., VIRTAMO J.T., "Optimal degree distribution for LT codes with small message length", *Proceedings IEEE Computer and Communications Societies (INFOCOMM)*, Anchorage, USA, IEEE, pp. 2576–2580, May 2007.

[IEE] IEEE standard for information technology part 15.4 – IEEE 802.15.4-2006, http://standards.ieee.org/getieee802/download/802.15.4-2006.pdf.

[JAF 09] JAFFRÈS-RUNSER K., COMANICIU C., GORCE J.-M., ZHANG R., "U-GRAB: A utility-based gradient broadcasting algorithm for wireless sensor networks", *Proceedings IEEE Military Communications Conference (MILCOM)*, pp. 1–7, October 2009.

[KAM 06] KAMRA A., MISRA V., FELDMAN J., RUBENSTEIN D., "Growth codes: maximizing sensor network data persistence", *Proceedings Conference on Applications, Technologies, Architectures, and Protocols for Computer Communications (SIGCOMM)*, New York, USA, pp. 255–266, 2006.

[KAR 04] KARP R., LUBY M., SHOKROLLAHI A., "Finite length analysis of LT codes", *Proceedings of International Symposium on Information Theory (ISIT)*, Chicago, USA, p. 39, 2004.

[KHA 02] KHANDEKAR A., Graph-based codes and iterative decoding, PhD thesis, California Institute of Technology, 2002.

[LIN 07a] LIN Y., LI B., LIANG B., "Differentiated data persistence with priority random linear codes", *Proceedings of the 27th International Conference on Distributed Computing Systems (ICDCS)*, p. 47, June 2007.

[LIN 07b] LIN Y., LIANG B., LI B., "Data persistence in large-scale sensor networks with decentralized fountain codes", *Proceedings of the IEEE International Conference on Computer Communications (INFOCOM)*, Anchorage, Alaska, USA, pp. 1658–1666, 2007.

[LU 09] LU F., FOH C.H., CAI J., CHIA L.-T., "LT codes decoding: Design and analysis", *Proceedings of International Symposium on Information Theory (ISIT)*, Seoul, South Korea, pp. 2492–2496, 2009.

[LUB 01] LUBY M., MITZENMACHER M., SHOKROLLAHI M.A., SPIELMAN D.A., "Efficient erasure correcting codes", *IEEE Transactions on Information Theory*, vol. 47, no. 2, pp. 569–584, 2001.

[LUK 09] LUKIC M., PAVKOVIC B., MITTON N., STOJMENOVIC I., "Greedy geographic routing algorithms in real environment", *Mobile Ad hoc and Sensor Networks, MSN '09, 5th International Conference on*, pp. 86–93, December 2009.

[MAA 09] MAATOUK G., SHOKROLLAHI A., "Analysis of the second moment of the LT decoder", *International Symposium on Information Theory (ISIT)*, Seoul, South Korea, pp. 2326–2330, 2009.

[MAC 05] MACKAY D.J.C., "Fountain codes", *IEE Communications*, vol. 152, pp. 1062–1068, 2005.

[MAN 06] MANEVA E., SHOKROLLAHI A., "New model for rigorous analysis of LT-codes", *Proceedings of International Symposium on Information Theory (ISIT)*, Seattle, USA, pp. 2677–2679, 2006.

[NOD 10] NODIN L., APAVATJRUT A., GOURSAUD C., GORCE J.-M., "Degree distribution of XORed fountain codes: Theoretical derivation and analysis", *Proceedings of Asia-Pacific Conference on Communications (APCC)*, Auckland, New Zealand, November 2010.

[OBR 98] OBRACZKA K., "Multicast transport protocols: a survey and taxonomy", *IEEE Communications Magazine*, vol. 36, no. 1, pp. 94–102, January 1998.

[PAK 07] PAKZAD P., SHOKROLLAHI A., "EXIT functions for LT and raptor codes, and asymptotic ranks of random matrices", *Proceedings of IEEE International Symposium on Information Theory (ISIT)*, Nice, France, pp. 411–415, 2007.

[PAU 97] PAUL S., SABNANI K., LIN J.-H., BHATTACHARYYA S., "Reliable multicast transport protocol (RMTP)", *IEEE Journal on Selected Areas in Communications*, vol. 15, no. 3, pp. 407–421, April 1997.

[POL 04] POLASTRE J., HILL J., CULLER D., "Versatile low power media access for wireless sensor networks", *Proceedings of the 2nd AMC International Conference on Embedded Networked Sensor Systems (SenSys)*, New York, USA, pp. 95–107, 2004.

[PUJ 04] PUJOLLE G., SALVATORI O., NOZICK J., *Les Réseaux*, Eyrolles, 2004.

[ROS 08] ROSSI M., ZANCA G., STABELLINI L., CREPALDI R., HARRIS A., ZORZI M., "SYNAPSE: A Network reprogramming protocol for wireless sensor networks using fountain codes", *Proceedings of IEEE Communications Society Conference on Sensor, Mesh and Ad Hoc Communications and Networks (SECON)*, San Francisco, USA, pp. 188–196, 2008.

[SEJ 09] SEJDINOVIC D., PIECHOCKI R., DOUFEXI A., ISMAIL M., "Fountain code design for data multicast with side information", *IEEE Transactions on Wireless Communications*, vol. 8, no. 10, pp. 5155–5165, 2009.

[SHA 48] SHANNON C.E., "A mathematical theory of communication", *Bell Systems Technical Journal*, vol. 27, pp. 379–423, 623–656, 1948.

[SHO 06] SHOKROLLAHI A., "Raptor codes", *IEEE Transactions on Information Theory*, vol. 52, no. 6, pp. 2551–2567, 2006.

[THO 08] THOMAS W., Energy-efficient self-organization for wireless sensor networks, PhD thesis, INSA-Lyon, 2008.

[TIR 09] TIRRONEN T., Fountain codes: performance analysis and optimisation, PhD thesis, Helsinki University of Technology, 2009.

[WSN] WSNet / Worldsens simulator, http://wsnet.gforge.inria.fr/.

[YE 03] YE F., ZHONG G., CHENG J., LU S., ZHANG L., "PEAS: A robust energy conserving protocol for long-lived sensor networks", *Proceedings of the 23rd International Conference on Distributed Computing Systems*, pp. 28–37, May 2003.

[YE 05] YE F., ZHONG G., LU S., ZHANG L., "Gradient broadcast: A robust data delivery protocol for large scale sensor networks", *Wireless Network*, vol. 11, pp. 285–298, May 2005.

Chapter 3

Switched Code for *Ad Hoc* Networks: Optimizing the Diffusion by Using Network Coding

3.1. Abstract

A wireless *ad hoc* network is made of a decentralized set of mobile and self-organized objects. Diffusion is one of the existing problems in this type of network. In such a communication, a message is sent from a given object toward all the other objects in the network. Most routing protocols use this process to diffuse their control messages. They are generally diffused by flooding (or pure diffusion), which can become very costly.

Network coding is a new technique that allows the routers to combine the fluxes they receive and to redirect these combinations toward different routes. It has been shown that network coding, combined with wireless diffusion, can

Chapter written by Nour KADI and Khaldoun AL AGHA.

improve the performances in terms of flow, energy efficiency, and bandwidth usage. Most existing approaches that use network coding are based on Galois corpora coding operations to combine the packets. As a consequence, they have a high calculation complexity. This chapter introduces the use of network coding to optimize the diffusions in *ad hoc* networks with the aim of finding a coding scheme with simple complexity while improving the flow.

The chapter shows how to decrease the decoding complexity by applying a new coding scheme that carries out operations on a binary corpus. Instead of combining the received packets linearly, it is sufficient to employ algorithmic sums modulo 2. The system then carries out coding and decoding with a logarithmic complexity.

At each transmission, a node chooses some packets to code by using a predefined distribution. To decrease the delivery period, a new distribution is introduced allowing the intermediate nodes to start decoding even if only a few coded packets are received.

3.2. Introduction

In the field of telecommunications networks, the research theme on *ad hoc* networks appeared about 15 years ago. The main feature of an *ad hoc* network is that a node can call for other nodes in its vicinity to retransmit data packets whose recipient is out of reach. Each participant can hence be both host and router. The management of these networks always raises interesting research problems in the fields of networks, distributed algorithmics, and digital communications. The use of radio resources, which are by nature particularly fluctuating and limited, constitutes a continually flourishing research theme. As a consequence, a type of distributed routing protocol is necessary to warrant, at each time, the connection between any pair of nodes in the network. Given the limits of *ad hoc* networks (dynamic topologies, a limited bandwidth, energy constraints, limited physical safety), the conventional routing schemes designed for wired networks are not appropriate in a mobile *ad hoc* environment. The *ad hoc* routing protocol must find the paths between the nodes with a minimum bandwidth consumption. Furthermore, it should have a low calculation complexity.

According to the method followed to establish communication routes, the routing protocols of *ad hoc* networks can be classified into two main categories: the proactive protocols (like optimized link state routing, OLSR) and reactive protocols (such as *ad hoc* on demand distance vector, AODV). Proactive protocols attempt to maintain in each node up-to-date routing information concerning all the other nodes of the network. This technique forces the nodes to flood the network with control packets. The reactive protocols search for a route in the network only when a node wishes to send a message. This search is carried out by flooding.

All the *ad hoc* routing protocols imply the flooding of control messages to find a path between two nodes of the network. The easiest way to diffuse a packet in the communication networks is by simple flooding or by flooding referred to as "blind" [OBR 01]. In this technique, a node retransmits to its neighbors all the packets it receives for the first time. Thus, nodes receive the same packet several times. Due to the simplicity of calculation, this solution is largely adopted at the level of wired networks. However, this is not appropriate in *ad hoc* networks because it provokes several problems such as collisions and significant energy consumption. As a consequence, the improvement in flooding efficiency has a very significant impact on the performance of these networks.

The network coding technique reduces the consumption of resources in a network. When we want to diffuse messages to the entire network by using this technique, the nodes of the network recombine the received message and then transmit a coded message.

By using the tools of information theory, the authors of [CHR 06] have shown that network coding can offer advantages during the application of flooding in wireless environments. The flooding with the help of network coding has been presented in several approaches [CHO 03, CHR 08, JOR 05, RUI 08]. Most of these approaches use linear network coding. In this technique, a node uses random linear combinations of available packets to perform the coding. Gaussian elimination is used to decode the source symbols. Due to their cubic calculation complexity, these approaches are expensive in terms of energy, processing, and required memory. Furthermore, a receiver must wait to receive a sufficient number of coded packets before recovering the source symbols.

To optimize the diffusion such as the one presented in [ERR 07], another type of network coding has been used in the literature. It is referred to as opportunistic network coding. This approach has a simple complexity because coding and decoding are carried out by using simple XOR operations. However, the efficiency of coding decisions depends on the exchange of a type of control information between neighbors. As a consequence, this approach increases the control overload.

From the above discussion, it seems important to find a coding scheme with a simple complexity, which enables us to optimize the flooding without introducing much delay. The following challenges must be taken into account for an efficient flooding protocol:

– Cost: the source packets must be given to the entire network with a minimum number of transmissions.

– Decentralization: the solution presented must be distributed, without any coordination or central data.

– Time limit: the time taken for the source packets to be delivered to the entire network should be minimum.

– Memory requirement: the amount of memory necessary to maintain the received packets for the retransmission, or the decoding, must be minimum.

– Processing time: the sending or the reception of packets must be done with a minimum amount of processing.

3.3. Diffusion in *ad hoc* networks

Diffusion is a configuration in which a node sends a packet to all the other nodes in the network. It plays an important role in several *ad hoc* routing protocols, such as AODV [ELI 99], dynamic source routing (DSR) [JOH 96], and OLSR protocol [CLA 01]. It is used to establish links between each source–destination pair or to discover the network topology. Other applications of diffusion in *ad hoc* wireless networks include service discovery, network management, information collection, and alarm sending.

The easiest method to carry out diffusion is known as blind flooding [OBR 01], which works as follows: the source node sends information to all its neighbor and the neighbors relay the messages that have been received for the first time to their neighbors. This re-diffusion takes place until all the network nodes receive a copy of the packet.

Blind flooding provokes a large number of redundant retransmissions, which is shown in Figure 3.1 where we notice that several nodes receive the same packet several times. This mechanism is very expensive in a wireless *ad hoc* network, because it consumes many resources (bandwidth and power, for instance). Furthermore, several neighbors can decide to retransmit the received packet at the same time. In this case, the radio interferences and collisions could break the network.

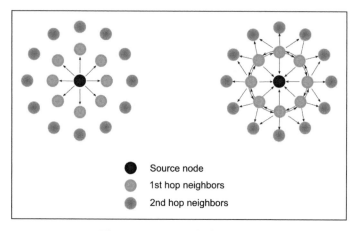

Figure 3.1. *Simple flooding*

To decrease the traffic generated by the diffusion of messages, a new technique has been suggested for OLSR protocol [CLA 01]. This technique is referred to as multipoints relays. It consists of selecting the minimum number of neighbors at one hop away, which enables us to reach all the neighbors at two hops away. Figure 3.2 clearly shows the advantage of using optimized flooding by using multipoints relays. In this figure, we notice that the number of retransmissions necessary to reach all the nodes of the network has been significantly decreased. To maintain all the up-to-date information necessary to the choice of multipoints relays, OLSR nodes need to exchange information periodically.

3.4. Diffusion and network coding

Network coding is a new area of research that has been presented by Ahlswede *et al.* in [AHL 00]. It has several interesting applications in wired and wireless networks.

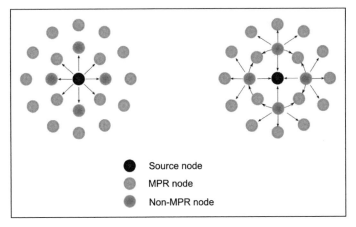

Figure 3.2. *MPR flooding*

Contrary to conventional routing in which the relay nodes cannot replicate the received packets, network coding allows the intermediate nodes to process the received information to generate a packet toward the exit. Several advantages can be obtained by applying network coding to communication networks, such as output gain, bandwidth sparing, and resistance against losses.

Let us take the example shown in Figure 3.3 to prove a case where network coding outperforms conventional routing for diffusion. In this network (Figure 3.3), there are two sources A and B that need to diffuse their information a and b, respectively, to the entire network. Each edge can carry only a single value. If we use conventional routing, Figure 3.3(a), then central relay R will need two transmissions to offer a and b to nodes C and D. With the use of a simple code, Figure 3.3(b), the relay of R delivers a and b to two destinations at the same time by sending the sum of symbols (in other words, it encodes a and b by using formula "$a \oplus b$"). Node C receives a and $a + b$, and we find b by XORing the two values. The same procedure is used in node D to find a. Thus, network coding decreases the number of transmissions required for the diffusion of source symbols.

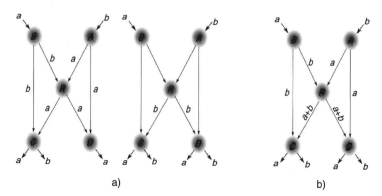

a)   b)

Figure 3.3. *Butterfly network: example of gain by using network coding*

Diffusion by using network coding has been presented in several approaches. These approaches can be divided into two categories: linear network coding and opportunistic network coding.

The approaches that use linear network coding [CHO 03, CHR 08, JOR 05] combine network coding with a probabilistic algorithm. In these approaches, a node uses random linear combinations of available packets to carry out the coding. The destination needs to receive a sufficient number of coded packets to be able to solve the equations and to decode the original messages. If it does not receive the required number of packets, then the decoding process fails and the destination will not be able to recover the source messages.

Let $x_1, x_2, \ldots, x_n$ be the n source packets to be delivered to the multiple destinations in a network with the help of network coding. These packets contain symbols on a corpus F_q and they can be generated by a single or several sources. With linear network coding, each packet x is associated with a vector of coefficients $\vec{c} = [c_1, c_2, \ldots, c_n]$ in F_q named a coding vector. It shows how packet x can be inferred from the source packets.

$$x = c_1 x_1 + c_2 x_2 + \cdots + c_n x_n$$

Let us consider a node that has received and stored m encoded packets $Y(1), Y(2), \ldots, Y(m)$. Each of these packets $Y(i)$ is associated with coding vector $\overrightarrow{g(i)}$. When this node wants to send a packet, it re-encodes the packets already coded as follows: it randomly chooses a set of coefficients $\overrightarrow{h} = [h_1, h_2, \ldots, h_m]$ and calculates the linear combination

$$Y' = h_1 Y(1) + h_2 Y(2) + \cdots + h_m Y(m).$$

The coding vector to be associated with this new coded packet is given by

$$\overrightarrow{g'} = h_1 \overrightarrow{g(1)} + h_2 \overrightarrow{g(2)} + \cdots + h_m \overrightarrow{g(m)}$$

A destination can recover the original packets if it receives n packets with linearly independent coding vectors, referred to as n innovating packets. Let $Y(i)$ and $\overrightarrow{g(i)}$ be the received coded packets and their coding vectors, respectively. Then, we can write a matrix of form

$$
\begin{bmatrix} Y(1) \\ Y(2) \\ \ldots \\ Y(n) \end{bmatrix} = \underbrace{\begin{bmatrix} g_1(1) & g_2(1) & \ldots & g_n(1) \\ g_1(2) & g_2(2) & \ldots & g_n(2) \\ \ldots & \ldots & \ldots & \ldots \\ g_1(n) & g_2(n) & \ldots & g_n(n) \end{bmatrix}}_{A} \begin{bmatrix} x_1 \\ x_2 \\ \ldots \\ x_n \end{bmatrix}
\qquad [3.1]
$$

If the n coding vectors are independent, then matrix A is invertible. Thus, the previous equation can be solved by Gaussian elimination and the source packets can be recovered.

Due to their high calculation complexity, these approaches are expensive in terms of energy, processing and memory requirements. Furthermore, a receiver must wait to receive a sufficient number of coded packets before recovering the source symbols. It is for this reason that these approaches could not be used in applications sensitive to delays because of the delay in decoding. In a network with losses, if some coded packets are lost, then the nodes would not receive sufficient packets for the decoding process, and it would not be able to recover a source symbol, which decreases the delivery rate of these approaches.

Opportunistic network coding has been used in [RUI 08, ERR 07, NOU 08]. These approaches require a weak calculation complexity. However, a node in these approaches needs a deep knowledge of the reception information of the neighbors to make a coding decision. In addition to the overload generated by the exchange of this information between the neighbors, this information cannot be constantly up-to-date, because of the mobility, loss or congestion.

The erasure codes have also been used to optimize diffusion. [ERR 07] suggests an algorithm to use Reed–Solomon coding. In this approach, a sender needs to transmit a set of k packets together, and its neighbors must receive all the packets to complete the decoding. If one packet of the set is lost, then all the received packets are wasted. The LT code [LUB 02] is another coding scheme that decreases the coding and decoding complexity. The idea behind the LT code is as follows: a source can generate an unlimited number of coded packets at a destination. For the coding, the source randomly and uniformly chooses d distinct symbols from k source symbols, XORs the set, and transmits it. d is referred to as the degree of the coded symbol and it is chosen according to a random distribution. The destination must receive n coded symbols to decode the source symbols. The efficiency of the LT codes lies in its degree distribution, which is referred to as robust soliton distribution (RSD). This distribution combines the ideal soliton distribution $\rho(.)$ and $\tau(.)$ where

$$\rho(i) = \begin{cases} \frac{1}{k} & i = 1 \\ \frac{1}{i(i-1)} & 1 < i \le k \end{cases}$$

$$\tau(i) = \begin{cases} R/ik & 1 \le i < \frac{k}{R} \\ \frac{R}{k} ln \frac{R}{\delta} & i = \frac{k}{R} \\ 0 & i > \frac{k}{R} \end{cases}$$

$$R = c.ln(\frac{k}{\delta})\sqrt{k}$$

k is the total number of source packets, $c > 0$ is a constant, and δ is the maximum probability of a decoding failure. In this approach, the intermediate nodes are only authorized to reproduce and transmit the packets, but do not take part in the coding process. According to [DES 05], this approach decreases the network flow and restricts the capacity to reach the min-cut limit. To avoid this drawback, Pakazad *et al.* in [PAY 05] present coding schemes that are based on the LT code. They enable an intermediate processing to reach the min-cut capacity. However, these schemes increase the complexity, the delay, and the required memory.

3.5. Switched code: incorporate erasure codes with network coding

Here, we consider that each node in the network generates packets to be diffused to all the other nodes. The *ad hoc* network is modeled by a unit disk graph. In such a graph, the radio range of a node is circular and of unit distance. Each transmission of a node x is received by all its neighbors $N(x)$, which are placed within a transmission radius of x. Our goal is to decrease the number of transmissions necessary to deliver the source packets in such a network and to enhance reliability and flow. For this reason, we suggest a switched code that carries out an efficient diffusion.

Switched code uses the following techniques:

– Combination of network coding and erasure code: each time an intermediate node has the possibility of sending, it generates a coded packet. This is formed by using an erasure code that is applied in a distributed way. The decoding process is similar to low density parity check (LDPC) decoding codes [MAC 03].

– Anticipated decoding: the code allows a node to decode any received packet immediately by using only the available packets. Any packet that is freed by this operation can take part later in the encoding or decoding operation.

3.5.1. *Definitions*

In the discussion, we refer to a native packet that is not coded as a *singleton*. The degree of a coded packet d is the number of native packets that are XORed together to form the packet, and we refer to these d packets as coding candidates.

Each node manages the following buffers: R-buffer is used to keep "during a certain time" the native packets whose node generates, receives or recovers (decodes) during the decoding process. B-buffer is used to keep the native packets to diffuse them. These packets can be generated by the node or can be coming from a neighbor. A packet is withdrawn from B-buffer when it is delivered to all the neighbors. E-buffer is used to keep the received coded packets, and these cannot be decoded immediately.

3.5.2. *Coding function of switched code*

Each time a node gains contention of the channel, it checks its B-buffer for the packets to be diffused. If B-buffer is not empty, then the encoding process takes place as follows:

– randomly choose a degree d of the coded packet according to a predefined distribution;

– randomly choose d distinct packets of B-buffer to use them as coding candidates;

– XOR the coding candidates to obtain the coded packet;

– add the ID of coding candidates in the header, then diffuse the coded packets.

The encoding process resembles the LT code, but it enables us to encode a node as well as the packets coming from different sources.

3.6. Decoding function of switched code

When a node receives a coded packet, it follows an online decoding process as shown in Figure 3.4. In contrast to the LT code that does not allow the receiver to start decoding until it receives a complete block of coded packets, switched code allows the receiver to carry out an anticipated decoding.

When a node receives a coded packet P that is made up of d native packets, first it attempts to decrease its degree. By using the ID of coding candidates present in the header, the node tries to recover the corresponding packets of R-buffer. If one of these packets is found, it is XORed with P to decrease its degree by one. If the previous step manages to decrease the degree of P to 1, then the new native packet P^* is decoded and inserted in R-buffer. P^* is used in a background decoding to decode other coded packets and stored in E-buffer. During background decoding, if the degree of a coded packet becomes 1, we repeat the same procedure for the new decoded packet. Any native packet freed during this process is inserted in B-buffer.

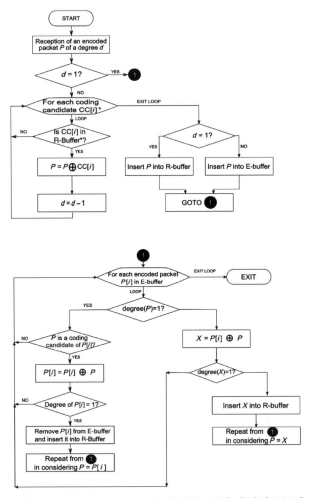

Figure 3.4. *Decoding process of switched code*

An example of the decoding process is shown in Figure 3.5. A node has three native packets in R-buffer – $p1, p2$ and $p3$ – and two encoded packets in E-buffer. It receives a new coded packet $p2 \oplus p3 \oplus p4$. First, it attempts to decode the received packets by using the native packets in R-buffer. The node

XORs the received packet with two coding candidates $p2$ and $p3$ to recover $p4$. As the recovered packet is a coding candidate of the second packet of E-buffer, they will be XORed together to decrease the degree of coded packets.

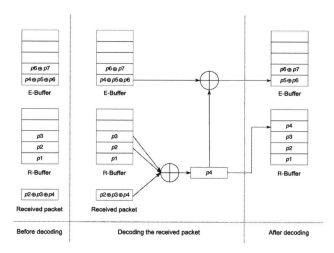

Figure 3.5. *Decoding example using switched code*

When a node sends a packet, it inserts in the packet header acknowledgment of receipt (ACKnowledgment) of recovered symbols. The receiver uses the discharge information to erase the packets of B-buffer when they are delivered to all the neighbors.

3.7. Design and analysis of a new distribution

In the previous section we saw that the encoding and decoding functions are similar to those of the LT code [LUB 02]. The LT code uses the RSD to generate the degree. As mentioned earlier, the LT code has been designed essentially for end-to-end coding. As a consequence, RSD is designed to decrease the general costs, but it allows the destination to start decoding only when it receives a sufficiently large number of coded symbols. This is illustrated in Figure 3.6, where we consider a source node that needs to send 1,000 packets toward a destination. The source and the destination are connected through a direct link without error. The source sends encoded packets generated according to RSD, and the destination uses the online decoding described in section 3.6. Parameters c and δ of RSD are equal to 0.2 and 0.1, respectively. We plot the number of symbols recovered at the destination as a function of the number of received coded packets. RSD generates coded packets of high degrees, which cannot be decoded until a sufficient number of source symbols are recovered. For this reason, we notice in Figure 3.6 that the reception of 1,200 packets allows the destination to recover only 0.3 of the source packets. Then, we observe the quick acceleration of the decoding process when the destination receives 1,300 packets.

In an *ad hoc* network, using the RSD distribution, the intermediate nodes can generate a huge delay, because the number of decoded packets at the level of these nodes can be modest at the beginning of the decoding process. As a consequence, we need a new degree distribution, which allows the decoded to recover (decode) sufficient symbols, even if only a few coded symbols are received.

To solve this problem, we suggest switched distribution. The idea underlying this distribution is to go from one distribution to the other according to the number of coded symbols that have been sent. At each time, the distribution that ensures a higher decoding probability is chosen. For instance, the receiver at the beginning of the communication process does not have sufficient singletons for these to be used during the decoding process. As a consequence, it would not be able to decode a packet of high degree at this time, since several of its coding candidates are not available. Decreasing the degree at the beginning of the communication process allows the receiver to decode the packet immediately by using only the singletons that are available locally. After a certain number of transmissions, it would be useful to increase the degree of coded packets because the receiver has already recovered several singletons.

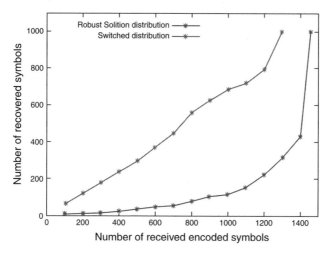

Figure 3.6. *Number of decoded packets at the destination as a function of the number of received packets;* $k = 1,000$

Switched code is defined as follows

$$\varpi_{i,k}(d) = \begin{cases} \varphi_k(d) & \text{for } i \le k \\ \mu_k(d) & \text{for } i > k \end{cases}$$

$$\varphi_k(d) = \begin{cases} \frac{1}{2^d} & \text{for } d = 1, 2, \ldots, k-1 \\ \frac{1}{2^{k-1}} & \text{for } d = k \end{cases}$$

where, $\varphi_k(d)$ is the binary exponential distribution (BED), $\mu_k(d)$ is the RSD, and k is the number of source packets.

Hence, according to $\varpi_{i,k}(d)$, the emitter uses BED to generate the ith encoded packet if i is smaller than k, and it uses RSD otherwise.

3.7.1. *Analysis of switched distribution*

In this section, we calculate the transition point between the two distributions used in switched distribution.

DEFINITION 3.1.– *(Codeword and degree): Codeword is the result of XORing several source symbols. These symbols are referred to as coding candidates. The number of coding candidates is referred to as codeword degree.*

DEFINITION 3.2.– BED_k *is given by*

- $\varphi(d) = \frac{1}{2^d}, \quad for \ \ d = 1, 2, \ldots, k-1$
- $\varphi(k) = \frac{1}{2^{k-1}}$

where k represents the total number of source symbols.

LEMMA 3.1.– *For all $k > 0$, BED_k is a probability distribution*

PROOF.–

$$\sum_{d=1}^{k} \varphi(d) = \sum_{1}^{k-1} \frac{1}{2^d} + \frac{1}{2^{k-1}} = 2 \times \left(1 - \frac{1}{2^k}\right) - 1 + \frac{1}{2^{k-1}} = 1$$

DEFINITION 3.3.– *(Decoding probability) Let $D_{(r|d)}$ be the probability of decoding a codeword of degree d where $r - 1$ of the source of symbols have been recovered.*

PROPOSITION 3.1.–

$$D_{(r|d)} = \begin{cases} (k - r + 1)/k & for \quad d = 1 \\ \dfrac{d.(k-r+1).\prod_{i=0}^{d-2}(r-1-i)}{\prod_{i=0}^{d-1}(k-i)} & for \quad d = 2, 3, \ldots, r \\ 0 & if \quad d > r \end{cases}$$

PROOF.– The destination, which has recovered $(r - 1)$ of symbols, is able to decode a received codeword of degree d if $(d - 1)$ of coding candidates are among $(r - 1)$ recovered symbols and a single candidate is among $(k-1r)$ non-recovered symbols. Hence, the decoding probability is

$$D_{(r|d)} = \frac{(k-r+1)\binom{r-1}{d-1}}{\binom{k}{d}} = \frac{(k-r+1).\frac{\prod_{i=0}^{d-2}(r-1-i)}{(d-1)!}}{\frac{\prod_{i=0}^{d-1}(k-i)}{d!}}$$

$$= \frac{d.(k-r+1).\prod_{i=0}^{d-2}(r-1-i)}{\prod_{i=0}^{d-1}(k-i)}$$

In the case where $d > r$, certainly more than one coding candidate are among the non-recovered symbols and, as a consequence it cannot be decoded.

DEFINITION 3.4.– *(The probability of recovery of a symbol) Let R_r be the probability of recovering r^{th} source symbol during the reception of a coded packet. Hence $R_r = \sum_{d=1}^{k} p(d).D_{(r|d)}$ where $p(d)$ is the probability of receiving a codeword of degree d.*

DEFINITION 3.5.– *Let E_y be the predicted number of recovered symbols after the sending of y codewords. And let $\Theta = Y - k$ be the overload where $E_Y = k$.*

In addition to minimizing the overhead, our interest is to maximize E_y, $\forall y$. This can be obtained by maximizing R_r, $\forall r$. Now, let us consider a source that has k packets to be sent to a destination through a direct link without loss.

LEMMA 3.2.– *For the destination to be able to recover the first symbol, it is more useful that the source chooses the degree of the codeword according to the BED_k than choosing the RSD.*

PROOF.– Before sending codewords, the number of recovered symbols at destination $(r-1)$ is 0. Only then can a codeword of degree 1 be decoded at this stage. In this case, the expected number of recovered symbols after the sending of the first codeword is

$$E_1 = p(d = 1) \times D_{(1|1)} = p(d = 1) \times 1$$

Let $p(d = 1) = \frac{1}{2}$ be the probability of obtaining a coded symbol of degree 1 when we use an exponential distribution, and $p'(d = 1) = (\frac{1}{k} + \frac{R}{k})/\beta$ represent the same probability during the use of RSD [LUB 02], where

$$R = c.ln(\frac{k}{\delta})\sqrt{k}$$

$$\beta = \sum_{i=1}^{\frac{k}{R}-1} \frac{R}{i} + R.ln\left(\frac{R}{\delta}\right) \leq 1 + \frac{R}{k}\left(H\left(\frac{k}{R}\right) + ln\left(\frac{R}{\delta}\right)\right)$$

and $H(n)$ is the number of harmonics of n.

We are going to prove that $p(d = 1) > p'(d = 1)$ is absurd.

Let us assume that

$$(\frac{1}{k} + \frac{R}{k})/\beta > \frac{1}{2}$$

$$\frac{1 + R}{k + R(H(\frac{k}{R}) + ln(\frac{R}{\delta}))} > \frac{1}{2}$$

$$k + R[H(\frac{k}{R}) + ln(\frac{R}{\delta}) - 2] < 2$$

This is impossible $\forall k > 1$. Hence, using BED_k increases the expected number of recovered symbols during the sending of the first coded symbol.

LEMMA 3.3.– *To recover the last symbol, it is easier to use the soliton distribution.*

PROOF.– We are going to prove the lemma for the ideal soliton distribution, and the results are valid for the robust distribution according to [LUB 02], where they prove that the probability of freeing a symbol by using the RSD is greater than this probability when we use the ideal soliton.

To recover the last symbol we have set r equal to k. According to proposition 3.1 we have $D_{(1|1)} = 1$ and $D_{(k|d)} = \frac{d}{k}$, $\forall k > 1$, $d \leq k$. Let us now compare the symbols recovery probability of both the distributions. According to definition 3.4, the symbols recovery probability $k > 1$ during the use of BED_k is

$$R_k = \frac{1}{k} \sum_{d=1}^{k} \frac{d}{2^d} = \frac{2}{k} \cdot \left[1 - \left(\frac{1}{2} \right)^{k+1} - \frac{k+1}{2^k} \right] \qquad [3.2]$$

and when we use the ideal soliton distribution it is

$$R'_k = \frac{1}{k} \left[\frac{1}{k} + \sum_{d=2}^{k} \frac{1}{d-1} \right] = \frac{1}{k} \left[\frac{1}{k} + \mathrm{H}(k-1) \right] \qquad [3.3]$$

$$R'_k - R_k = \frac{1}{k} \cdot \left[\frac{1}{k} + \mathrm{H}(k-1) - 2 + \frac{1}{2^k} + \frac{2k+2}{2^k} \right]$$

$$= \frac{1}{k} \cdot \left[\frac{1}{k} + \mathrm{H}(k-1) + \frac{2k+3}{2^k} - 2 \right] \qquad [3.4]$$

When $k = 1$, $p(1) = 1$ for both distributions. Then $R_1 = R'_1 = 1$.

When $2 \leq k \leq 4$, according to equation [3.4] we notice that $R_1 - R'_1 > 0$.

When $k \geq 5$, as $\mathrm{H}(k-1) > 2$ then $R_1 - R'_1 > 0$.

Thus, we conclude that $R'_k \geq R_k$, $\forall k > 0$.

LEMMA 3.4.– *DSR outperforms BED_k only after the recovery of 70% of source packets at the destination.*

PROOF.– Let R'_r and R''_r be the probability of recovery of a symbol using RSD and BED_k, respectively. Let

$$\Gamma_d = \frac{\prod_{i=0}^{d-2}(r-1-i)}{\prod_{i=0}^{d-1}(k-i)}$$

Then, we have

$$R'_r = \frac{k-r+1}{\beta}\left[\frac{R+1}{K^2} + \sum_{d=2}^{\frac{k}{R}-1}(\frac{1}{d-1} + \frac{R}{k}).\Gamma_d\right.$$

$$\left. + (\frac{R^2}{k(k-R)} + \frac{Rln(\frac{R}{\delta})}{k})d.\Gamma_{\frac{k}{R}} + \sum_{d=\frac{k}{R}+1}^{k}\frac{1}{(d-1)}\Gamma_d\right]$$

$$R''_r = (k-r+1).\sum_{d=1}^{r}\frac{d}{2^d}.\Gamma_d$$

By using a dichotomous technique, we notice that when $r-1$ is smaller than $0.70 \times k$, then R'' is greater than R' and it is the opposite when $r-1$ becomes greater than $0.70 \times k$.

To find the transition point between the two distributions, we need to estimate the number of codewords that the destination needs to recover 70% of source packets when we use BED_k. For this, we define an additional decoder that decodes the packets with restricted conditions.

DEFINITION 3.6.– *Let us take an incremental decoder S. If decoder S receives r codewords, then it decodes them in increasing order according to their degree d. In the first step, it decodes the packets with d = 1. Then, by using the packets that are recovered from the first step, it decodes the packets with d = 2. For a step i, it decodes the packets with d = i by using the packets that are recovered from steps 1, 2, . . . , i − 1. If a packet of degree d cannot be decoded at step i = d, then it will be disregarded in the following steps.*

PROPOSITION 3.2.– The expected number of recovered symbols after sending codewords k according to BED_k is at least $0.70 \times k$.

PROOF.– Let us prove this proposition first for the incremental decoder. Let us assume that S receives Y codewords, then:

– Step 1: All the codewords of degree 1 are decoded and their coding candidates are recovered. The expected number of codewords of degree 1 that could be decoded is $E_Y[1] = Y \times \varphi(d = 1) \times 1 = \frac{Y}{2}$

– Step 2: The expected number of codewords of degree i that could be decoded is

$$E_Y[i] = Y \times \varphi(d = i) \times D_{((\sum_{j=1}^{i-1} E_Y[j]+1)|i)}$$

$$= \frac{Y}{2^i} \times D_{((\sum_{j=1}^{i-1} E_Y[j]+1)|i)}$$

Then, the total expected number of recovered symbols is

$$E_Y = \sum_{i=1}^{k} E_Y[i]$$

$$= \frac{Y}{2} + \sum_{i=2}^{k} Y \times \varphi(d = i) \times D_{((\sum_{j=1}^{i-1} E_Y[j]+1)|i)}$$

$$= \frac{Y}{2} + \sum_{i=2}^{k} \frac{Y}{2^i} \frac{i.(k - \sum_{j=1}^{i-1} E_Y[j]).\prod_{q=0}^{i-2}(\sum_{j=1}^{i-1} E_Y[j] - q)}{\prod_{q=0}^{i-1}(k - q)}$$

By using a dichotomous technique, we notice that $Y = k$, then $E_Y = 0.70 \times k$. This means that when we use decoder S, the destination needs at least k codewords to recover 70% of k source symbols. This result is valid for a non-incremental decoder[1] because the number of recovered symbols during the

---

1 The non-incremental decoder decodes the codewords as they arrive without limitation linked to their order, and it keeps, for a later decoding, codewords that cannot be decoded immediately.

unlimited use of this decoder is greater than or equal to that of decoder S.

As a conclusion, the source sends k words by using BED_k to ensure the delivery of at least 70% of k source symbols at the destination, such as mentioned in proposition 3.2. Then, it switches to the RSD to deliver the remaining symbols because RSD increases the decoding probability at this stage.

3.8. Conclusion

In recent years, wireless networks have experienced a rapid growth due to the increase in popularity of wireless devices. A wireless *ad hoc* network is a set of devices with a weak battery power and a modest calculation power. They can communicate without the need of any infrastructure. A large number of challenges can be found in this type of network, such as limited bandwidth, limited transmission rate, and the extremely dynamic network topology.

The diffusion in *ad hoc* networks plays a leading role in most *ad hoc* routing protocols. This is the process in which a node sends a packet to all the other nodes of the network. Due to the features of *ad hoc* networks, this configuration has been challenged in these environments. Thus, efficient solutions for the diffusion are sought.

One such popular solution presented in the literature is network coding. It is a recently emerging field that has received a great deal of attention. It allows the nodes to process the input information, before transmitting it to its neighbors. It has been proved that network coding has the potential to significantly decrease the delay and the power consumption while improving the use of the bandwidth and the global robustness of the network.

The existing approaches that use network coding for diffusion in wireless *ad hoc* networks require a large calculation complexity. To avoid this drawback, we have presented a distributed coding scheme, switched code, which combines the techniques of rateless code and network coding. The use of rateless code, like the LT code, enables us to carry out coding and decoding with linear complexity. However, the LT code could increase the delay in transmission of end-to-end packets because of its high delay in decoding. We have decreased this delay by suggesting and analyzing a new distribution degree that allows the intermediate nodes to decode the packets as soon as they are received.

3.9. Bibliography

[AHL 00] AHLSWEDE R., CAI N., LI S., YEUNG R.W., "Network information flow", *IEEE Transactions on Information Theory*, vol. 46, pp. 1204–1216, 2000.

[CHO 03] CHOU P.A., WU Y., JAIN K., "Practical network coding", *Allerton Conference on Communication, Control, and Computing*, Monticello, USA, 2003.

[CHR 06] CHRISTINA F., JORG W., "On the benefits of network coding for wireless applications", *4th International Symposium on Modeling and Optimization in Mobile, Ad Hoc and Wireless Networks*, pp. 1–6, 2006.

[CHR 08] CHRISTINA F., JORG W., JEAN-YVES L.B., "Efficient broadcasting using network coding", *IEEE/ACM Transactions on Networking*, vol. 16, pp. 450–463, 2008.

[CLA 01] CLAUSEN T., QAYUUM A., JACQUET P., LAOUITTI A., VIENNOT L., "Optimized link state Routing protocol", *IEEE INMIC*, 2001.

[DES 05] DESMOND S.L., MURIEL M., RALF K., MICHELLE E., "On coding for reliable communication over packet networks", *CoRR*, vol. abs/cs/0510070, 2005.

[ELI 99] ELIZABETH M.R., CHARLES E.P., SAMIR R.D., "*Ad hoc on demand distance vector (AODV) routing*", *WMCSA*, IEEE Computer Society, 1999.

[ERR 07] ERRAN L.L., RAMACHANDRAN R., MILIND M.B., SCOTT C.M., "Network coding-based broadcast in mobile *ad hoc* networks", *INFOCOM*, IEEE, pp. 1739–1747, 2007.

[JOH 96] JOHNSON D.B., MALTZ D.A., "Dynamic source routing in *ad hoc* wireless networks", *Mobile Computing*, Kluwer Academic Publishers, pp. 153–181, 1996.

[JOR 05] JORG W., JEAN-YVES L.B., "Network coding for efficient communication in extreme networks", *Workshop on Delay Tolerant Networking and Related Networks (WDTN-05)*, New York, USA, vol. 05, 2005.

[LUB 02] LUBY M. "LT codes", *FOCS: IEEE Symposium on Foundations of Computer Science (FOCS)*, Vancouver, Canada, pp. 271–280, 2002.

[MAC 03] MACKAY D., *Information Theory, Inference, and Learning Algorithms*, Cambridge Univ. Pr., 2003.

[NOU 08] NOUR K., KHALDOUN A.A., "Optimized MPR-based flooding in wireless *ad hoc* network using network coding", *IFIP / IEEE Wireless Days'08*, Dubai, UAE, 2008.

[OBR 01] OBRACZKA K., VISWANATH K., TSUDIK G., "Flooding for reliable multicast in multi-hop *ad hoc* networks", *Wireless Network*, vol. 7, no. 6, pp. 627–634, 2001.

[PAY 05] PAYAM P., CHRISTINA F., AMIN S., "Coding schemes for line networks", *CoRR*, vol. abs/cs/0508124, 2005.

[RUI 08] RUI A.C., DANIELE M., JORG W., JOÃO B., "Informed network coding for minimum decoding delay", *CoRR*, vol. abs/0809.2152, 2008.

Chapter 4

Security by Network Coding

4.1. Introduction

Network coding is a highly useful technique because it allows us to asymptotically reach the capacity bounds to disseminate a message in a network. Due to the way it functions, network coding is an attractive method in terms of security. Today, there are at least 50 different research articles examining the use of network coding as a security mechanism.

The literature examined in this chapter focuses on security methods taken from information theory rather than from cryptology. The difference between the attacker models used in these two fields is highly important. This is because both the fields focus unconditionally on security but with different attacker models. In terms of cryptology, Shannon has shown that *one-time-pad* coding has an unconditional security, irrespective of an attacker's calculation power (and some unavoidable hypotheses on *one-time-pad* instantiation).

Chapter written by Katia JAFFRÈS-RUNSER and Cédric LAURADOUX.

At the same time, work by Wyner on *wiretap channels* has shown that it is possible to attain unconditional security for a coding system where only the attacker has access to a degraded version of the channel between the emitter and the proper receiver of a message. This is a strong hypothesis and limits the practical applications of results from the *wiretap channel*.

Here, we focus purely on the results linked to *"wiretap network"* in relation to Wyner's work. After identifying our attacker's capacities, we study code conception methods for *"wiretap networks"*. Following this, we examine an algebraic security criterion for random linear network coding that allows us to guarantee privacy.

4.2. Attack models

The attacks that we are going to examine in this chapter focus exclusively on the confidentiality of messages exchanged on the network on the basis of transmissions about which an attacker is trying to find information conveyed by source(s). It should be noted here that some research in information theory (see [JAG 05a, JAG 07]) has examined examples of byzantine adversaries. An attacker compromises an intermediary node and injects errors into it. If the damaging capacity of such an attacker, that is the number of errors injected, is limited, then the classic error correction methods can be employed. Following this, we focus on privacy with the issue of error injections (also known as *pollution*) examined in the following chapter.

Without an adequate measure or additional hypothesis, network coding does not guarantee security: an adversary capable of listening to enough transmissions is able to eventually decode the information. For network coding to guarantee confidentiality, two elements are necessary:

– limit the attacker's listening ability;

– to find the "good codes" for network coding.

The first element is resolved using the hypothesis employed in the first type II Ozarov and Wyner *wiretap channel* [OZA 84]. The attacker can only receive part of the information exchanged between the source and its destination. Figure 4.1 shows the type-II *wiretap channel* described by Ozarov and Wyner [OZA 84]. A *wiretap channel* can be seen as the simplest form of network. We use this to illustrate the attacker model used in information theory.

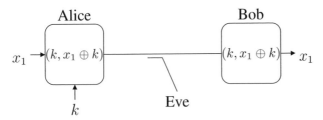

Figure 4.1. *The type-II wiretap channel*

The attacker, Eve, does not recover all the symbols transmitted by Alice to Bob. Alice needs to secretly send $K = 1$ bits of information to Bob, written as x_1. To do so, Alice takes a distributed random uniform bit k and calculates the bit: $x_1 \oplus k$ with $\oplus$, the modulo addition 2. Alice sends these $N = 2$ bits to Bob. If Eve can only see $\mu = 1$ bits transmitted by Alice, then this system has unconditional security. This is the same as that of the *one-time-pad* (see [SHA 49]). The type-II *wiretap channel* study consists of finding the codes that allow us to guarantee unconditional security for the different parameters K, N and μ.

It should be noted that the method employed in Figure 4.1 can be interpreted in two different ways. We can see it as a *one-time-pad* coding or as an application of secret sharing

[BLA 17, SHA 79]. This last interpretation is particularly interesting because it allows us to see the MDS codes resurface [MCE 81]. Approaching network coding security as a question of secret sharing is not, however, easy. This is because secret sharing is the complimentary problem of confidentiality in network coding. In a secret sharing scheme, we distribute parts of a secret to different participants from a Λ authorization list. The Λ list identifies the groups of participants who could reconstruct the secret. In the case of network coding confidentiality, we identify the entities that cannot recover the secret.

The type-II *wiretap channel* generalization to more complex networks was initiated by Cai and Yeung [CAI 02] in 2002 and the term *wiretap network* was first used by Rouayheb and Oljanin [ROU 07]. Having examined in further detail the attacker model for the *wiretap network*, we will introduce a restriction of this model called a *nice but curious* attacker [LIM 07].

4.2.1. *A type-II wiretap network*

We can illustrate a passive attack for the butterfly network in Figure 4.2. Alice wants to transmit her message x to Bob 1 and Bob 2. She splits the message into two packets x_1 and x_2 which she transmits to the relays Eve 1 and Eve 2, respectively. Eve 3 combines the messages received using a simple XOR addition and retransmits the message $x_1 + x_2$ to Eve 4, which transmits it to the destinations Bob 1 and Bob 2.

For an example of message overhearing on network links, the network code presented is secure if the attacker can only listen to one link at a time. However, if he or she is capable of listening to two lines simultaneously, he or she can, for the most part, decode the message.

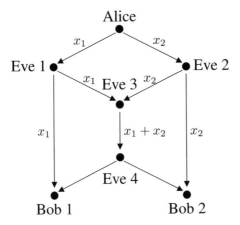

Figure 4.2. *A butterfly network*

To describe a *wiretap network*, we will use the symbols used by [CAI 11]. A *wiretap network* is, therefore, a quadruplet $(\mathcal{G}, s, \mathcal{U}, \mathcal{A})$ made up of the following components:

1) *The directed multi-graph $\mathcal{G}$*: we use the symbol $\mathcal{G} = (\mathcal{V}, \mathcal{E})$ to define a directed multi-graph, or $\mathcal{V}$ and $\mathcal{E}$ are the sets of nodes and edges of the graph $\mathcal{G}$, respectively.

2) *Source node s*: the set $\mathcal{V}$ contains a source node s that sends a message M generated on the basis of an alphabet $\mathcal{M}$.

3) *Set of user nodes $\mathcal{U}$*: a user node of $\mathcal{V}$ is completely accessible to an honest user who receives the message M without error. Normally, we use $\mathcal{U}$ to indicate the set of user nodes.

4) *Set of wiretap edges $\mathcal{A}$*: $\mathcal{A}$ is a collection of subsets of edges $\mathcal{E}$. Each element of $\mathcal{A}$ can be completely accessed by an attacker but no attacker can access more than one element of $\mathcal{A}$.

Cooperation between different attackers is a large threat for confidentiality based on network coding. In Figure 4.2, network coding can be considered safe against an attacker listening to a single edge. However, if we consider the collection of packets encoded on several links (e.g. on the link between Eve 3 and Eve 4 and on the link between Eve 2 and Bob 2) and it is possible to decode the transmitted messages using the formula $x_1 = (x_1 \oplus x_2) \oplus x_2$.

In summary, we can say that a network coding mechanism is said to be safe against a *wiretap network* if the attacker cannot obtain any information on the transmitted message without altering decoding by legitimate users.

4.2.2. *A nice but curious attacker*

A "nice but curious" attacker is found inside the network. He or she, therefore, has access to internal resources of nodes and its main objective is to listen and decode packets he or she sees communicated. The *nice but curious* label was introduced in research by Lima *et al.* in [LIM 07]. This kind of node uses operations conforming to the network protocol and only seeks to reduce the network's performance.

Figure 4.3 illustrates a *nice but curious* attack. In this example, Alice sends a message $m = (a_1, a_2, a_3, a_4, b_1, b_2, b_3, b_4)$ by dividing it into two streams of four packets $a = (a_1, a_2, a_3, a_4)$ and $b = (b_1, b_2, b_3, b_4)$. The node E_3 combines them to prevent subsequent relays in the network (here E_4 and E_5) and shows all the information transmitted by Alice. In this example, E_4 can reconstruct b knowing a and c, where c is a linear combination of the packets of a and b. Equally, E_5 can reconstruct a knowing b and d, where d is another linear combination of a and b. As such, E_4 and E_5 can understand the original message m.

To secure the network against *nice but curious* attackers, the principle idea proposed in [LIM 07] is to define an *algebraic security criterion* that guarantees that a random linear combination of data to the intermediary nodes prevents these nodes from understanding all the information. This is examined in further detail in section 4.4.

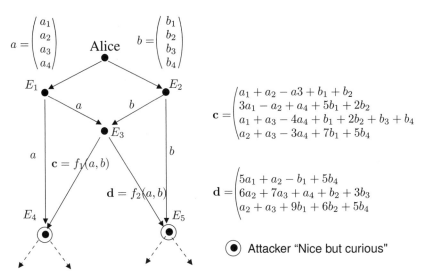

$$a = \begin{pmatrix} a_1 \\ a_2 \\ a_3 \\ a_4 \end{pmatrix} \qquad b = \begin{pmatrix} b_1 \\ b_2 \\ b_3 \\ b_4 \end{pmatrix}$$

$$c = \begin{vmatrix} a_1 + a_2 - a3 + b_1 + b_2 \\ 3a_1 - a_2 + a_4 + 5b_1 + 2b_2 \\ a_1 + a_3 - 4a_4 + b_1 + 2b_2 + b_3 + b_4 \\ a_2 + a_3 - 3a_4 + 7b_1 + 5b_4 \end{vmatrix}$$

$$d = \begin{vmatrix} 5a_1 + a_2 - b_1 + 5b_4 \\ 6a_2 + 7a_3 + a_4 + b_2 + 3b_3 \\ a_2 + a_3 + 9b_1 + 6b_2 + 5b_4 \end{vmatrix}$$

Attacker "Nice but curious"

Figure 4.3. *Example of network coding for a nice but curious attack on the nodes E_4 and E_5*

4.3. Security for a *wiretap network*

Cai and Yeung have defined a secure network code for a *wiretap network* in [CAI 11].

DEFINITION 4.1.– *A code is said to be secure for a wiretap network if it satisfies the following conditions:*

1) Decoding: $\forall u \in U$ *and* $\forall m, m' \in M$ *so that* $m \neq m'$

$$\Phi_u(m, k) \neq \Phi_u(m', k'), \qquad [4.1]$$

for every $k, k' \in \mathcal{K}$ *knowing that* k *and* k' *can be different and with* Φ_u *the function induced by network coding for the edge* u.

2) Security: $\forall W \in \mathcal{W}$

$$H(M|Y_W) = H(M). \qquad [4.2]$$

Here, $H(\cdot|\cdot)$ and $H(\cdot)$ are conditional entropy and entropy, respectively.

The condition on decoding guarantees that two distinct messages remain distinguishable for every node $u \in \mathcal{U}$. There is no couple $k, k' \in \mathcal{K}$ so that two distinct messages M and M' cause two identical transmission for the users u. The security condition stipulates that the information obtained by the attacker does not reduce the uncertainty of the message.

There is a large amount of literature focusing on researching the construction of a secure code for network coding. [CAI 11] studies conditions for linear codes and [ROU 07] examines an approach based on secret sharing. The main challenge is finding constraints on the size of the alphabet used by the source. Rouayheb and Soljanin have provided a fairly simple bound for the size of the alphabet taken from the *Linear Information Flow* (LIF) [JAG 05b] algorithm in [ROU 07].

THEOREM 4.1.– *Take a network $\mathcal{G} = (\mathcal{V}, \mathcal{E})$ with a single source, t destinations and min-cut between source and destination of n. For an attacker who can see $\mu \leq n - k$ edges with a flow $k < n$, we can still see a secure network code if the source employs an alphabet $\mathbb{F}_q$ of size*

$$q > \binom{|E| - 1}{\mu - 1} + t. \qquad [4.3]$$

We can again see the parallel between network coding and secret sharing. In both the disciplines we want to minimize the size of the data to be shared.

4.4. Algebraic security criteria

This section introduces the algebraic security criterion proposed by Lima, Médard, and Barros in [LIM 07]. This criterion aims to measure the local security level (i.e. at the level of a node provided intrisincally by the random linear network coding). In the first subsection we will examine the characteristics of random linear network coding. We will then introduce the notion of algebraic security defined in [LIM 07], and will follow this up by definining the algebraic security criterion.

4.4.1. *Note on random linear network coding*

Algebraic security has been defined in the framework of *random linear network coding*, for a multicast transmission of K symbols transmitted by one or several sources toward $d \geq 1$ destinations. Transmissions from all sources must be received by each destination. Each source transmits a series of K symbols modeled by carrying out K stochastic processes $X_1, X_2, \ldots, X_K$. Each process has a constant entropy of one bit by time unity and the connections in the network also have a capacity of one bit by time unity. The connections in the network are considered to be perfect: transmission is carried out without loss and no delays are introduced.

The network is modeled using a *directed acyclic graph* $G = (V, E)$ where V denotes the set of nodes in the network and E the set of edges. Each edge is denoted by $e = (v, u) \in E$ where the node $v = \text{head}(e)$ is the head and $u = \text{tail}(e)$ the tail of the directed arc. For each node $v \in V$, we define:

– $\Gamma_I(v)$ and $\Gamma_O(v)$ as the set of edges entering and leaving v;

– $\delta_I(v)$ and $\delta_O(v)$ as the inbound and outbound degree of edges entering and leaving v, respectively.

In a linear network coding [HO 06], each link $e(v, u)$ transports the combination of symbols, which are either generated by v if v is a source or transmitted. We can represent the state of the link by the process $Y(e)$:

$$Y(e) = \sum_{l:X_l \text{ is generated in } v} \alpha_{l,e} X(v,l) + \sum_{e':\text{head}(e')=\text{tail}(e)} \beta_{e',e} Y(e')$$

where, if v is a source:

- $X(v,l)$ are the symbols generated by v;

- $\alpha_{l,e}$ are the coefficients applied to the symbols generated by v and transmitted on e;

and where:

- $Y(e')$ is the process carried by an edge entering e',

- $\beta_{e',e}$ are the coefficients applied to the retransmitted symbols from e' and transmitted on e.

The coefficients α and β are chosen at random uniformly and independently among all the elements of a finished field $\mathbb{F}_q$, $q = 2^n$.

From the previous formulation, we can deduce a matricial representation. The *transfer matrix* $\mathbf{M}$ provides the linear combination between the vector of K symbols transmitted $\underline{x}$ and the vector received at one of the destinations $\underline{z} : \underline{z} = \underline{x}.\mathbf{M}$. The matrix $\mathbf{M}$ is split into $\mathbf{M} = \mathbf{A}(I - \mathbf{F})^{-1}\mathbf{B}^T$, where $\mathbf{A}$ and $\mathbf{B}$ represent the linear combinations of symbols on entry and the symbols on exit. $\mathbf{A}$ has size $K \times |E|$ and $\mathbf{B}$ has size $\nu \times |E|$. $\mathbf{F}$ is the adjacency matrix of the directed labeled graph taken from G.

If we have $\mathbf{G} = (I - \mathbf{F})^{-1}$, of size $|E| \times |E|$, we obtain $\mathbf{M} = \mathbf{AGB}^T$. To be able to decode $\underline{x}$ from $\underline{z}$, the matrix $\mathbf{M}$ has to be of full rank. Ho *et al.* have shown in [HO 06] that random linear coding provides a solution if the size q of the finite field $\mathbb{F}_q$ verifies if $q > d$, where d is the number of destinations. It is in this context that Lima *et al.* have defined algebraic security in [LIM 07].

4.4.2. *Algebraic security*

Algebraic security allows us to protect against an internal attack of the *nice but curious* nodes. This aims to limit the decoding ability of the intermediary nodes in the network, while guaranteeing the proper transmission of the message toward its destinations.

Security in the sense of information theory, introduced by Shannon in [SHA 49], corresponds to the example where the clear text M and the coded text C have zero mutual information: $I(M;C) = 0$. In this case, an attacker needs to decipher $H(M)$ symbols at most to access the entire message, with $H(M)$ being the entropy of the message M. In the case of network coding, if the source transmits K symbols that are recombined in the network, the attacker who has succeeded in solving m coded messages will again need $K - m$ messages to obtain the K original symbols. In order for an attacker to obtain a single symbol in network coding, he or she needs K encoded symbols. Figure 4.4 illustrates this example. In the upper section, attackers 2 and 3 intercept and can understand most of the messages. In the lower section, attackers 2 and 3 cannot understand any intercepted message by the combination of symbols created by node 1.

4.4.3. *The algebraic security criterion*

This criterion is measured by the number of symbols that an intermediary node v needs to predict to be able to decode a transmitted symbol. An intermediary node $v \in V$ of the network sees a certain number of coded symbols pass on the network. The symbols that it sees on the channel are determined by the transfer matrix **M** and, more specifically, by the entries of the matrix **AG** that correspond to the subset of symbols that the node v has seen pass on the channel.

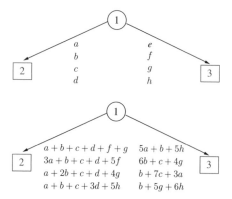

Figure 4.4. *Example of algebraic security. In the upper figure, the symbols are not protected and attackers 2 and 3 can understand most of the symbols transmitted. However, in the lower part, 2 and 3 can only recover the messages transmitted by 1*

Formally, we can define the *partial transfer matrix* $\mathbf{M}'_{\Gamma_I(v)} = \mathbf{AG}_{\gamma_I(v)}$ that gives the value of the coefficients applied to K transmitted symbols and that the node v passes. $\mathbf{M}'_{\Gamma_I(v)}$ is given by the columns of the matrix $\mathbf{AG}$ corresponding to the entry edges $\Gamma_I(v)$. It is of the dimension $K \times \delta_I(v)$. This allows us to deduce a fraction of the information transmitted by S to which the node v has access.

DEFINITION 4.2.– *The security level $\Delta_S(v)$ provided by a random linear network coding is measured by the number of symbols that a node v has to predict to decode a symbol transmitted by the source. It is expressed by*

$$\Delta_S(v) = \frac{K - \left(rang(\mathbf{M}'_{\Gamma_I(v)}) + l_d \right)}{K} \qquad [4.4]$$

where l_d represents the number of lines of $\mathbf{M}'_{\Gamma_I(v)}$ that are partially diagonalizable, that is the number of original symbols which can be decoded by a Gaussian elimination.

This definition is equivalent to the calculation of the difference between the global rank of the network code **AG** and the rank observed at an intermediary node v.

It has been shown in [LIM 07] that the probability that an attacker can obtain $X < K - 1$ symbols tends toward 0, irrespective of its entry degree verifying $\delta_I(v) < K$, if the size of the field q and the number of symbols K tend toward infinity. In other words, the security level of a random linear network code is as guaranteed as the field size and number of symbols is large for the intermediary nodes of degree $\delta_I(v) < K$.

In the case where $\delta_I(v) \geq K$, it is possible that v sees K independent linear combinations and as such obtains a matrix $M'_{\Gamma_I(v)}$ that is equal to **AG** (i.e. a full rank matrix). This situation does not always necessarily occur. Its occurrence depends on the type of network and the distribution of flow in the network. According to the type of graph G, some nodes at higher entry degree at K can only receive combinations of symbols from partial matrices of lower rank at K.

4.4.4. *Algorithmic application of the criterion*

The algebraic security criterion proposed by [LIM 07] is introduced as secure for the random linear network coding using a field size and a number of arbitrarily large packets, for one of the nodes that present a lower entry degree at K. In order for the nodes to have a higher inbound degree of K, it is necessary to modify the network's structure to be sure that the rank of the partial matrix remains less than or equal to $K - 1$ in these nodes.

It is in this sense that the authors of [GEN 08] propose a distributed algorithm that allows us to construct a multipath transmission using a random linear network code between a source and a receiver in a sensor network, which are secured in the sense of the algebraic criterion of Lima *et al*. They consider that all the relays are *nice but curious* attackers. The authors seek to protect the integrity of the data transmitted by a source toward a destination based on the *directed diffusion* routing algorithm, one of the reference algorithms for routing in sensor networks.

4.5. Conclusion

In this chapter, we have studied means of using network coding to protect the confidentiality of communication against different attacker models. We have examined an example of network coding with a single source. The issue of confidentiality in the context of multiple sources has been introduced in [CHA 08]. The main difficulty in ensuring confidentiality by network coding is being able to guarantee the conditions of the type-II *wiretap network*. Without these hypotheses, network coding does not allow us to protect messages. If we cannot limit the listening ability of the attacker, we have to employ more conventional cryptographic techniques that will be examined in the following chapter.

4.6. Bibliography

[BLA 17] BLAKLEY G.R., "Safeguarding cryptographic keys", *AFIPS 1979 National Computer Conference*, vol. 48, Arlington, NY, USA, pp. 313–317, 1979.

[CAI 02] CAI N., YEUNG R., "Secure network coding", *IEEE International Symposium on Information Theory – ISIT 2002*, pp. 323–329, June 2002.

[CAI 11] CAI N., YEUNG R.W., "Secure network coding on a wiretap network", *IEEE Transactions on Information Theory*, vol. 57, no. 1, pp. 424–435, 2011.

[CHA 08] CHAN T., GRANT A., "Capacity bounds for secure network coding", *Australian Communications Theory Workshop, 2008–AusCTW 2008*, IEEE, pp. 95–100, February 2008.

[GEN 08] GENG L., LU F., LIANG Y.-C., CHIN F., "Secure multi-path construction in wireless sensor networks using network coding", *International Symposium on Personal, Indoor and Mobile Radio Communications – PIMRC 2008*, pp. 1–5, September 2008.

[HO 06] HO T., MÉDARD M., KOETTER R., KARGER D.R., EFFROS M., SHI J., LEONG B., "A random linear network coding approach to multicast", *IEEE Transactions on Information Theory*, vol. 52, no. 10, pp. 4413–4430, October 2006.

[JAG 05a] JAGGI S., LANGBERG M., HO T., EFFROS M., "Correction of adversarial errors in networks", *Proceedings of the 2005 International Symposium on Information Theory – ISIT 2005*, pp. 1455–1459, September 2005.

[JAG 05b] JAGGI S., SANDERS P., CHOU P.A., EFFROS M., EGNER S., JAIN K., TOLHUIZEN L.M.G.M., "Polynomial time algorithms for multicast network code construction", *IEEE Transactions on Information Theory*, vol. 51, no. 6, pp. 1973–1982, 2005.

[JAG 07] JAGGI S., LANGBERG M., HO T., KATABI D., MÉDARD M., "Resilient network coding in the presence of byzantine adversaries", *INFOCOM*, IEEE, pp. 616–624, 2007.

[LIM 07] LIMA L., MEDARD M., BARROS J., "Random linear network coding: A free cipher?", *IEEE International Symposium on Information Theory – ISIT 2007*, pp. 546–550, June 2007.

[MCE 81] MCELIECE R.J., SARWATE D.V., "On sharing secrets and Reed-Solomon codes", *Communication of the ACM*, vol. 24, no. 9, pp. 583–584, ACM, 1981.

[OZA 84] OZAROW L.H., WYNER A.D., "Wire-Tap Channel II", *Advances in Cryptology – EUROCRYPT 84*, Lecture Notes in Computer Science 209, Springer, pp. 33–50, April 1984.

[ROU 07] ROUAYHEB S.Y.E., SOLJANIN E., "On wiretap networks II", *IEEE International Symposium on Information Theory – ISIT 2007*, pp. 551–555, June 2007.

Chapter 5

Security for Network Coding

5.1. Introduction

In this chapter, we will be interested in solutions suggested and especially developed to warrant the security of network coding. These solutions become necessary when the level of security intrinsically given by network coding is no longer sufficient. This chapter presents security solutions designed to ward off attacks specific to network coding. Being a new communication technique, network coding leads to new attacks. Hence, new security solutions specific to network coding have been developed and are presented in this chapter. Among the properties of conventional security that must be fulfilled in a secured network, we are going to be interested here in two fundamental properties: confidentiality and integrity/authenticity.

This chapter is organized as follows. In section 5.2, we give a taxonomy of attacks that exist against network

Chapter written by Marine MINIER, Yuanyuan ZHANG and Wassim ZNAÏDI.

coding by using a classification based on the power of the attacker (computer/sensor), the nature of the attack (active/passive), and the place of the attacker in the network (internal/external). From this threat model, confidentiality and authenticity are tackled, respectively, in sections 5.3 and 5.4.

Section 5.3 presents the conditions that must fulfill the confidentiality mechanisms if we want to use them for network coding. These schemes are illustrated with the help of practical examples. In section 5.4, we formalize and give examples of integrity and authenticity solutions dedicated to network coding. These solutions enable us to protect ourselves against a very dangerous attack and one which is in particular energy consuming: pollution attack.

5.2. Attack models

Several threats hang over the current systems of network coding. In [DON 09b], the authors have classified the adversaries into two categories: *external adversary* and *internal adversary*. The adversary that can listen passively to the network, and inject, modify, or replay (i.e. re-emit) the packets between nodes is called external adversary. It cannot access network resources, nor have access to internal states of nodes. The internal adversary, on the other hand, has at its disposal more capabilities. It can "disguise" itself in one or several legitimate participant(s). The internal adversaries are in general nodes that have been compromised. Hence, the internal adversary has access to internal data of compromised nodes: it has at its disposal some keys. We will, however, not discuss here the full meaning of the term internal adversary and how it is able to compromise the equipment and data of the nodes.

We provide in this section a finer characterization of attack models, which are *passive attacks* and *active attacks*.

5.2.1. *Eavesdroppers*

An "eavesdropper" does not perturb the good working of a network, but it tries to learn or to make use of the information exchanged in the network without altering information. It ensues therefrom that confidentiality can be violated if an attacker is also able to interpret the data collected because of the listening. Usually, the aim is not to detect passive attacks, but to prevent an intelligible listening. The ciphering of data constitutes a sufficient measure against a passive attacker.

Passive listening can come from inside the network or from outside. In the case of an internal attacker, it is seen as a legitimate node, while, in the case of an external attacker, it can only collect the data transmitted between the nodes.

5.2.1.1. *Internal eavesdroppers*

The case of internal eavesdropper has been dealt with partly in the previous chapter with the "nice but curious" attacker. The use of ciphering must be combined with mechanisms of key distributions to limit the impact of internal eavesdroppers. In fact, if a single ciphering key is shared by all the network nodes then confidentiality cannot be warranted against an eavesdropper. By using a system of key distribution to establish point-to-point secured links between nodes, we bring the problem of internal eavesdropper back to a "nice but curious" attacker issue. We obviously cannot disregard the security of the mechanism of key distribution.

5.2.1.2. *External eavesdroppers*

Within the framework of wiretap channel (see the previous chapter), we have studied the case of an external eavesdropper having a limited listening ability. However, this hypothesis is not satisfactory in several applications. In this chapter, we study the case of an attacker who has at its disposal an unlimited listening ability. This means that it has at its disposal means to listen to all the communications existing in the network. It can be a league of adversaries of a single entity, which is able to listen to all the network.

5.2.2. *Active attackers*

The main difference between an eavesdropper and an active attacker is that the attacker acts on the system. Its aim is to alter the working of the network by modifying, for instance the messages. Two active attacks have been studied in particular: the pollution attack and the flooding attack. The difference between an internal and an external attacker is paramount for active attacks. We can deal with the case of an external active attacker but not with that of an internal active attacker. It is easy enough to see that a corrupted node can generate incorrect data by fulfilling the terms of security protocols put in place. In what follows, we consider the case of only the external attacker in our description.

5.2.2.1. *Pollution attacks*

One of the most dangerous attacks against network coding is referred to as pollution attack [DON 09a, DON 09b]. These attacks are well known in the field of peer-to-peer networks and have naturally found an echo in the field of network coding. The pollution attacks exploit the ability of network coding to diffuse information for mischievous purposes. An adversary who corrupts some communications in the network can manage to corrupt all the results generated by network coding. This type of attack clearly has an epidemic aspect.

Let us now consider the communication scenario described in Figure 5.1 with an adversary Charlie and no security policy.

Charlie can corrupt message x_1 exchanged between Alice and Eve. Eve receives $x_1' = x_1 \oplus \epsilon$ instead of x_1. After having received x_2, Eve diffuses $x_1 \oplus x_2 \oplus \epsilon$ instead of $x_1 \oplus x_2$ to Alice and Bob. At the end of communications, Alice and Bob are, respectively, going to obtain $x_2' = x_2 \oplus \epsilon$ and $x_1' = x_1 \oplus \epsilon$. These messages are not the messages that Alice and Bob should receive. Charlie uses network coding to its advantage and makes sure that the pollution of a single message has an effect via the amplifying role of Eve on the messages received by both Alice and Bob.

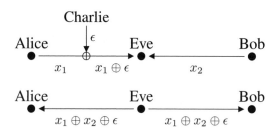

Figure 5.1. *An example of pollution attack*

To mitigate this attack, the role of the relay node, here Eve, is critical. Several strategies have been suggested in the literature to prevent pollution attacks in network coding, such as the use of homomorphic message authentication code (MAC) and homomorphic signatures. The methods relative to the pollution attack are detailed in section 5.4.

5.2.2.2. *Flooding attack*

Flooding attacks [DON 09b] are the attacks that focus on acknowledgment (ACK) packets.

In a random network coding mechanism, the sending of information stops only when the source node receives an ACK packet coming from one or several recipient nodes indicating that the message has been properly decoded. Flooding attacks are very easy to put in place in wireless networks due to the nature of the radio medium. To launch a flooding attack, an adversary can generate, delete or delay the ACK. The consequences are either that the transmission of information can stop (generation of ACK) or that the set of nodes of the network carries on the transmission even if it is completed.

In this section, we are going to present the results of the domain enabling us to protect data confidentiality when network coding itself is not able to supply such a security property. In this case, we have to deal with an attacker who is an external eavesdropper and whose aim is to listen to the packets and to attack data confidentiality. A simple and common way to warrant data confidentiality is to cipher it.

Similar to network coding where the intermediate nodes can calculate linear combinations of received packets, the ciphering methods used to protect data confidentiality must have specific properties to ensure end-to-end confidentiality (i.e. confidentiality must be warranted between the two final nodes). One of the most useful properties enabling us to fulfill these conditions is the property of homomorphism.

In this section, we first present the properties of homomorphism and complete homomorphism, and then we present a few homomorphic ciphering systems dedicated to network coding. Finally, we present the solutions from the literature that do not rely on homomorphic ciphering and can also be used to protect data confidentiality.

5.2.3. *Definition of homomorphic ciphering schemes*

As defined in [FON 07], a ciphering scheme E of a set M of plaintexts toward a set C of ciphertexts is referred to as homomorphic if for any given ciphering key k, E fulfills

$$\forall m_1, m_2 \in M, \quad E(m_1 \odot_M m_2) \leftarrow E(m_1) \odot_C E(m_2)$$

for operators $\odot_M$ of M and $\odot_C$ of C where the notation "$\leftarrow$" means "can be directly calculated from" (i.e. without intermediate deciphering).

This definition is really general, and most of the time homomorphic ciphering schemes are useful if the operators $\odot_M$ and $\odot_C$ are group or field operators such as $+$ and $\times$. Hence we can refine the aforementioned definition in two ways:

– A ciphering scheme is referred to as additively homomorphic if

$$\forall m_1, m_2 \in M, \quad E(m_1 +_M m_2) \leftarrow E(m_1) +_C E(m_2).$$

– A ciphering scheme is referred to as multiplicatively homomorphic if

$$\forall m_1, m_2 \in M, \quad E(m_1 \times_M m_2) \leftarrow E(m_1) \times_C E(m_2).$$

Several ciphering schemes are either additive or multiplicative. For a complete study on homomorphic schemes, the reader can refer to [FON 07]. Among the most well-known schemes, we can cite Goldwasser–Micali's scheme [GOL 82], which uses an RSA number and Jacobi symbols calculations, and Paillier's scheme [PAI 99], which is additively homomorphic. However, Goldwasser–Micali's scheme is not very efficient because to cipher 1 bit, it must generate a corresponding ciphering of size 1,024 bits. These two schemes and many others are based on the same principles as public key cryptography and the conventional and difficult problems of number theory. There also exists a simple additively homomorphic ciphering, which is based only on primitives of symmetrical cryptography: it is one-time pad cipher [BAU 05]. In what follows, we detail the Paillier's scheme and one-time pad cipher. Paillier's scheme has been used in [FAN 09], for instance to preserve confidentiality of packets exchanged during a network coding process.

5.2.3.1. *Two specific schemes*

5.2.3.1.1. Paillier' scheme

Paillier's scheme works as follows:

– Choose two large prime numbers p and q, keep them secret. Calculate and make public $n = pq$, the resulting modulo of 1,024 bits. Choose and make public g so that $\gcd(L(g^\lambda \pmod{n})^2), n) = 1$ with $L(u) = \frac{u-1}{n}$ and $\lambda(n) = \mathrm{lcm}(p-1, q-1)$.

– Cipher the plaintext $m \in \mathbb{Z}_n$ as follows: $c = g^m r^n (\pmod{n})^2)$ with r randomly chosen in $\mathbb{Z}_n$. Send c.

– The deciphering process is the following: $m = \dfrac{L(c^\lambda \pmod{n})^2)}{L(g^\lambda \pmod{n})^2)}$.

Although the use of homomorphic ciphering schemes is an elegant solution, it is extremely expensive: the size of the message ciphered by Paillier is about 2,048 bits and the complexity of the algorithm is also high and requires many calculations.

5.2.3.1.2. One-time pad ciphering

In one-time pad ciphering, the two entities that want to securely communicate must first share a common ciphering suite k. Once this exchange is securely carried out, when Alice wants to send a message to Bob, she simply XORs the plaintext with the ciphering suite. To decipher, Bob simply calculates the XOR between the received ciphertext and the ciphering suite, which is the same for Alice and Bob.

This scheme has been proved unconditionally safe by Shannon if and only if:

– the ciphering suite is really random;

– the ciphering suite has the same size as the plaintext; and

– the ciphering suite is used once and only once.

What is really used in everyday life is a modified version of the original scheme where the ciphering suite is generated with the help of an initialized stream cipher algorithm with a secret symmetrical key shared by Alice and Bob and a public value of initialization denoted IV, which changes at each transmission. Some examples of stream ciphers are SNOW v2.0 [EKD 02], SOSEMANUK [BER 08], or obviously AES in counting mode [NAT 01], etc.

We must also mention that the elementary operation used here ($\oplus$ on $\{0, 1\} = \mathbb{F}_2$) can be replaced by any bijective operator, typically a group law $(G, +)$.

A last important remark concerns the cost in terms of calculations of a stream cipher in particular if we compare it to any public key ciphering scheme: there is a gain in terms of speed in the order of 1,000 in favor of the family of stream ciphers because the latter uses only very efficient elementary operations both in hardware and in software.

5.2.3.2. *Completely homomorphic encryption schemes*

In 2009, in [GEN 09], Gentry introduced a more powerful concept than that defined previously and named as a completely homomorphic encryption. This scheme enables us to evaluate circuits on ciphered data without calling for deciphering. In other words, a completely homomorphic ciphering scheme allows any person to publicly transform a collection of ciphertexts from plaintexts $\pi_1, \ldots, \pi_n$ into a ciphertext for a function/circuit $f(\pi_1, \ldots, \pi_n)$ of plaintexts, without the person in question needing to know the plaintexts. Clearly, this concept is stronger than the previously defined "simple" homomorphic property.

From this definition, the author in [GEN 09] presents first a general result enabling us to build such completely homomorphic primitives. It then describes a possible instantiation of this definition based on a scheme built from ideal Euclidean networks. Unfortunately, this scheme cannot be used in practice because of its very high complexity in terms of calculations.

Since that seminal work, other suggestions have been made, essentially based on Euclidean networks. We can cite here [SMA 10], where the size of ciphertexts and the key sizes have been decreased; [DIJ 10], which suggests a scheme based on additions and multiplications on integer sets; and [STE 10], which improves Gentry's analysis.

This research area is relatively new and is hence yet to develop. However, the completely homomorphic ciphering schemes suggested in the literature often remain theoretical and so expensive in terms of calculation time and in terms of required parameter lengths that they cannot be used at the moment. The development of this area leads us to think that soon we will have at our disposal a really usable completely homomorphic encryption scheme.

5.2.4. *Homomorphic encryption and confidentiality in network coding*

Now that we have given the definitions of homomorphic and completely homomorphic encryption schemes, we are going to be interested in their use enabling us to warrant end-to-end confidentiality in network coding.

This section is divided in two sub-sections. The first deals with the case of network coding using XOR. In this case, a unique homomorphic property is required because only an XOR operator is used by all the nodes. The second sub-section is devoted to the more general case of network coding using for instance random linear codes or equivalent codes. In this second case, as the intermediate nodes can calculate linear combinations of received packets before retransmitting them, the completely homomorphic encryption property can be elegantly implemented.

5.2.4.1. *The case of network coding using XOR*

In the rest of this section, our analysis focuses on the simplest case of network coding using XOR in intra-flow as presented in Figure 5.2. Let us assume that Alice wants to send Bob a message p broken down into two fragments p_1 and p_2. For this, Alice broadcasts p_1 and p_2, and Charlie listens to both fragments and transfers to Bob $p_1 \oplus p_2$. As Bob has heard p_1, from the information sent by Charlie, he is able to decode p_2 and hence to recover the complete message. Clearly, in this simple case, network coding only does not enable us to prevent Charlie from understanding all the message p. Thus, a ciphering process must be added to protect data confidentiality.

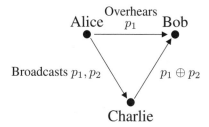

Figure 5.2. *A simple example of network coding using an intra-flow XOR*

The public key homomorphic encryption schemes (like Paillier's) can be used to cipher each fragments as soon as Bob will know that it is Alice who sends the fragments and that he will have knowledge of Alice's public key. For that purpose, Alice ciphers with her secret key KsA each fragment before broadcasting it: $c_1 = E_{KsA}(p_1)$ and $c_2 = E_{KsA}(p_2)$. Bob hears c_1 and receives from Charlie $c_1 \oplus c_2$. Then Bob deciphers c_1 as $p_1 = D_{KpA}(c_1)$ where KpA is Alice's public key. Similarly, he deciphers $c_1 \oplus c_2 = D_{KpA}(p_1 \oplus p_2)$, which is possible because encryption is homomorphic, and calculates p_2.

Unfortunately, because of the length of the cipher required by public key cryptography (in general at least 1,024 bits for conventional public key cryptography and 160 bits for cryptography based on elliptical curves), this form of cryptography is poorly adapted to the requirements of network coding.

A simple solution to solve the problem of confidentiality in network coding using an intra-flow XOR is the aforementioned one-time pad cipher. The ciphered suite is generated from a stream cipher algorithm initialized with a public value IV and a secret key k shared between the two entities wishing to communicate. This scenario applied to our example works in the following way: first, Alice and Bob need to build a shared key k (we will not detail here this step, we just say that such construction methods exist). Alice then generates two ciphered suites r_1 and r_2 from two different IVs IV_1 and IV_2 and from key k. It then ciphers p_1 and p_2 with $c_1 = p_1 \oplus r_1$ and $c_2 = p_2 \oplus r_2$ and then it broadcasts (c_1, IV_1) and (c_2, IV_2). (Let us note that in this case, the IV used to cipher the fragments must be linked to him to allow the destination to properly decipher the fragments received.) Bob hears (c_1, IV_1) and $(c_1 \oplus c_2, IV_1, IV_2)$. From c_1, Bob calculates p_1 and from $c_1 \oplus c_2$ he calculates $p_1 \oplus p_2$ and hence p_2. As here, the length of ciphertexts is equal to the length of plaintexts; this solution seems to be better adapted when small message fragments are considered. However, as the IVs must be transmitted for the proper deciphering, the size of each fragment is no longer constant but linearly depends on the number of fragments.

A last remark must also be mentioned about the example chosen. In fact, this is too simple to directly convince the reader of the necessity of using a homomorphic cipher: it is not clear if it is necessary to decipher p_1 then $p_1 \oplus p_2$. However, it is easy to be convinced of the usefulness of such an encryption as soon as we consider at least three fragments p_i: in this case, Bob must start the deciphering on-the-fly for each packet received before starting decoding.

Unfortunately, the solutions suggested here cannot be applied directly to the case of inter-flow network coding, where the intermediate nodes combine the traffic coming from several sources toward one (or several) destination, because this process would become too complex. First public key cryptography can no longer be used in this case, the destination is no longer able to decipher fragments of the form: $c_1 = E_{KsA}(p_1) \oplus E_{KsC}(p_2)$, the homomorphic property is then useless. However, the one-time pad cipher can still be used because it has an interesting additional property:

$$\forall m_1, m_2 \in M, E_{k_1 \oplus k_2}(m_1 \oplus m_2) \leftarrow E_{k_1}(m_1) \oplus E_{k_2}(m_2). \quad [5.1]$$

However, in the case of the use of one-time pad cipher for inter-flow network coding, the identity of each node emitting fragment must be transmitted in the fragments recombining several fragments coming from different nodes. This is the same for each value of IV, which leads to a strong increase in the size of the fragments to be transmitted. This also implies that the advantage of one-time pad cipher becomes extremely restricted in the case of inter-flow network coding.

5.2.4.2. *The case of network coding in general*

If we consider the more general case of network coding, such as the use of random linear codes, the simple homomorphic property used in the previous section is not sufficient to properly cipher and decipher packets. However, a potential solution clearly remains the use of completely homomorphic encryption schemes, where the function f considered becomes a linear combination, if the codes used are random linear for instance. Let us consider two fragments p_1 and p_2 of a message p belonging to a group G possessing an additive law $+$ and a scalar product $\times$. The completely homomorphic encryption scheme E_k defined according to our needs will have to fulfill $c_1 = E_K(p_1)$ and $c_2 = E_K(p_2)$, $\alpha_1 \times c_1 + \alpha_2 \times c_2 = E_K(\alpha_1 \times p_1 + \alpha_2 \times p_2)$.

Unfortunately, currently, no really efficient scheme has been suggested in the literature apart from those based on bilinear couplings on elliptical curves (such as those described in [BON 07]) and by considering a restricted definition of function f. Furthermore, these schemes can be used only for intra-flow network coding.

The case of completely homomorphic encryption schemes, which would work for inter-flow network coding (i.e. schemes that would additionally possess the property given by equation [5.1]), remains an open problem and it is not totally clear whether it is possible for cryptographs to build such schemes in light of the conventional security properties required.

5.3. Confidentiality

5.3.1. *Alternatives for confidentiality*

To our knowledge, there exists only one article [ZHA 10] that attempts to solve the problem of confidentiality preservation in data exchanged through network coding. The suggested method uses "lightweight" permutations to cipher, which are applied to each message and to its coding vector. The name of this suggestion is P-coding.

Let us assume that a source node s wants to send h packets $X = x_1, \cdots, x_h$ to a set of confidence sinks. Each packet x_i is defined as a line vector $x_i = [x_{i,1}, \cdots, x_{i,l}]$ of length l, defined on a finite field $\mathbb{F}$. Then, by assuming that if $\Gamma^-(s)$ represents the set of links ending in s, made of h imaginary links $e_1, \cdots, e_h$, with $y(e_i) = \mathbf{x}_i$, then, for all $e \in \Gamma^+(v)$ (the set of links starting from v), $y(e)$ is calculated as the combination of input packets of v, linearly:

$$y(e) = \sum_{e' \in \Gamma^-(v)} \beta_{e'}(e) y(e') = \beta(e) Y(e')$$

where coefficients $\beta_{e'}$ are randomly chosen in $\mathbb{F}$ and form vector $\beta(e) = [\beta_{e'}(e)]$.

By recurrence, $y(e)$ can be represented as a linear combination of elements of X:

$$y(e) = \sum_{i=1}^{h} g_i(e) \mathbf{x}_i = \mathbf{g}(e) X$$

where $\mathbf{g}(e) = [g_1(e), \cdots, g_h(e)]$.

By assuming that h packets $y(e_1), \cdots, y(e_h)$ are received by the sinks, we can rewrite the previous equation in matrix form as

$$Y = [y(e_1), \cdots, y(e_h)]^T = [\mathbf{g}(e_1), \cdots, \mathbf{g}(e_h)]^T X = GX$$

If G, known under the name of global encoding matrix (GEM), is invertible, then node v can decode X by using G^{-1}.

In reality, the source prefixes each packet $\mathbf{x}_i$ by i^{th} unit vector $\mathbf{u}_i$:

$$[\mathbf{u}_i, \mathbf{x}_i] = [0, \cdots, 0, 1, 0, \cdots, 0, x_{i,1}, \cdots, x_{i,l}]$$

The suggested ciphering scheme is hence the following: let $\mathbf{m} = [m_1, \cdots, m_n]$ be a symbol sequence of $\mathbb{F}$, then the permutation encryption function (PEF) on $\mathbf{m}$ using key k is defined as:

$$E_k(\mathbf{m}) = E_k([m_1, \cdots, m_n]) = [m_{k(1)}, \cdots, m_{k(n)}].$$

The deciphering process is defined by:

$$D_k(\mathbf{c}) = D_k([c_1, \cdots, c_n]) = [c_{k^{-1}(1)}, \cdots, c_{k^{-1}(n)}].$$

This function has two specific properties:
– addition: $E_k(\mathbf{m} + \mathbf{n}) = E_k(\mathbf{m}) + E_k(\mathbf{n})$;
– scalar product: $E_k(t \times \mathbf{m}) = t \times E_k(\mathbf{m})$.

Cipher E_k is a permutation built from a shared key k by using the algorithm of random permutation generation described in [KNU 73].

The suggested P-coding scheme hence consists of applying ciphering function E_k previously defined to each coded message $y(e_i)$ to obtain the corresponding ciphertext: $c[y(e_i)] = E_k[y(e_i)]$. Source node s sends its vectors, and the intermediate nodes recode until a sufficient number of messages have reached sink v. Because of the properties of the ciphering function, the sink can decipher the message $c[y(e_i)]$: $D_k\{c[y(e_i)]\} = E_{k^{-1}}\{E_k[y(e_i)]\} = y(e_i)$ and hence, by using the matrix representation, decode X from the received packets.

However, if many fragments of data are delivered by using the same key, certain bits of the key can be exposed. It is hence necessary to add a process enabling us to perturb the key. The authors also provide security proofs of their solution and show how the random feature is well preserved. As a conclusion, this scheme seems really efficient in terms of transmission because the network overload induced by this method is zero-valued.

5.4. Integrity and authenticity solutions

In this section, we are going to present the existing solutions enabling us to solve the pollution attack problem: if network routers are harmful and transfer invalid combinations of received packets, then these invalid packets are going to be mixed with valid packets downstream and are quickly going to pollute all the network. Furthermore, the receiver who obtains several packets has no means of discerning the valid packets that must be used for decoding. In fact, using as little as one polluted packet during the decoding would prevent any valid decoding.

To prevent the network from being completely flooded with polluted packets, it is necessary to have "hop-by-hop confinement". More precisely, this means that a polluted packet injected in the network must be detected and put aside at the following hop to be eliminated before being combined with other packets. This solution enables us to prevent the pollution from contaminating all the network.

As described in [AGR 09, BON 09], hop-by-hop confinement cannot be carried out with the help of signatures or of MAC standards. In fact, the source simply cannot sign or authenticate the packets to be sent in the network (by using conventional schemes), because the packets at the destination will clearly be different (they will in fact be linear combinations) from original packets. The destination will therefore not be able to check the signatures of the MACs.

Furthermore, signing or authenticating the whole message before its transmission does not work either because in this case the destination must first decode the received packets and again find the message itself before being able to check the signature. Thus, if the intermediate packets are polluted, the destination must decode any possible subset of a sufficient dimension of packets and hope to find again the correct message and only then check the signature.

Thus, new integrity and authenticity mechanisms must be put in place to protect oneself against pollution attacks. This section is organized as follows: section 5.4.1 gives the formal definitions of homomorphic MAC and homomorphic hash functions. It also presents some existing solutions based on these definitions. Section 5.4.2 gives a formal definition of the signature for network coding and presents some existing solutions. Finally, section 5.4.3 is dedicated to other solutions that are not based on the previous definitions.

5.4.1. *Definitions of homomorphic MAC and homomorphic hash functions*

5.4.1.1. *Definition*

The following definition is a modified version of the definition presented in [LI 10], whose first version was published in [AGR 09].

DEFINITION 5.1.– *A homomorphic MAC must fulfill the following properties [AGR 09, LI 10]:*

*1) **Homomorphism.** Given two pairs (message,tag) $(\mathbf{m}_1, t_1)$ and $(\mathbf{m}_2, t_2)$, anyone can create a valid tag t_a for an aggregated message $\mathbf{m}_a = \omega_1 \mathbf{m}_1 + \omega_2 \mathbf{m}_2$ for any scalar ω_1 and ω_2. Typically, $t_a = \omega_1 t_1 + \omega_2 t_2$.*

*2) **Security against chosen-plaintext attacks.** Even under a chosen-plaintext attack, in which the adversary is authorized to ask for a number of tags that is a polynomial function of the number of messages, it is always impossible for this adversary to create a valid tag for a message other than a linear combination of messages previously asked.*

*A homomorphic MAC consists of three probabilistic algorithms polynomial in time (*Sign, Combine, Verify*):*

– $t_u = $ Sign$(k, \mathrm{rid}, \mathbf{m}_u, \mathrm{id}_u)$: node u with ID id_u, contributor of message $\mathbf{m}_u$ concerning the report rid, calculates a tag t_u for $\mathbf{m}_u$ by using k as key.

– $t = $ Combine$((\mathbf{m}_1, t_1, \omega_1), \cdots, (\mathbf{m}_n, t_n, \omega_n))$: a combiner implements the homomorphic property for the pair

(message,tag) in the absence of key k, in other words, it generates tag t for combined message $\mathbf{m} = \sum_{i=1}^{n} \omega_i \mathbf{m}_i$.

– Verify(k, rid, $\mathbf{m}$, t): a verifier checks the integrity of the message $\mathbf{m}$ in view of report rid with the help of key k and tag t.

As mentioned earlier, this definition was first introduced in [AGR 09] and first constructions were also suggested based on the universal hash functions suggested originally by Carter and Wegman in [CAR 79]. These constructions use a pseudo-random generator and a pseudo-random function. Another scheme based on two pseudo-random generators has also been suggested in [LI 10].

DEFINITION 5.2.– *A homomorphic hash function H is a hash function that fulfills [LI 10, DIJ 10, KRO 04]:*

*1) **Homomorphism.** For any message pair $\mathbf{m}_1$, $\mathbf{m}_2$ and any scalar pair ω_1, ω_2, we must have $H(\omega_1 \mathbf{m}_1 + \omega_2 \mathbf{m}_2) = H(\mathbf{m}_1)^{\omega_1} H(\mathbf{m}_2)^{\omega_2}$ (let us note that the group operation considered here is the multiplication).*

*2) **Resistance to collisions.** There does not exist any probabilistic adversary polynomial in time (PPT) able to forge $(\mathbf{m}_1, \mathbf{m}_2, \mathbf{m}_3, \omega_1, \omega_2)$ fulfilling both $\mathbf{m}_3 \neq \omega_1 \mathbf{m}_1 + \omega_2 \mathbf{m}_2$ and $H(\mathbf{m}_3) = H(\mathbf{m}_1)^{\omega_1} H(\mathbf{m}_2)^{\omega_2}$.*

5.4.1.2. *Examples of such schemes*

The first two schemes suggested in the literature were simple MAC additively homomorphic. They were suggested by Krawczyk in 1994 [KRA 94]. These schemes can be directly used to prevent pollution attacks in network coding using XOR as suggested in [ANY 10] where all the possible checking policies for intermediate nodes are also described.

As mentioned earlier, the homomorphic schemes have been described in [LI 10] and [AGR 09]. In [AGR 09], after having suggested a first construction, the authors extend it to a homomorphic MAC where broadcasting is possible by using a technique described by Canetti *et al.* in [CAN 99]. This allows all the nodes of the network to validate the vectors it receives. Unfortunately, this scheme is resistant only against c nodes, which cooperate where c is a predefined value. The authors finally modify the former MAC suggestion into an integrity scheme authorizing multiple senders and receivers.

Homomorphic hash functions fulfilling definition 5.2 can be used in several fields such as securization of network coding against pollution attacks as suggested in [DIJ 10] and the securization of peer-to-peer contents by using erasure codes [KRO 04]. Other examples of homomorphic hash functions are given in [GKA 05] and [GKA 06], but the suggested two schemes are not efficient in terms of calculation cost and in over-communication induced for the network. In [LI 10], the authors suggest two other homomorphic hash function schemes. The methods used to build such functions are the discrete logarithm and integer factorization.

5.4.2. *Definition of homomorphic signature schemes*

As previously explained, MACs rely on secret key cryptography algorithms. Hence they require the use of a unique key that must be shared between all the checkers or the use of several key pairs. To solve this problem, the use of public key cryptography can be a solution leading to define homomorphic signature schemes. In this section, we will first give this definition before presenting examples of these schemes.

5.4.2.1. *Definition*

DEFINITION 5.3.– *A signature scheme for network coding [BON 09] is a quadruplet of probabilistic algorithms*

*polynomial in time (Setup, Sign, Combine, Verify) having
the following features:*

   – Setup($1^k, N$). *This algorithm takes as inputs a security
parameter 1^k and an integer N. It supplies as outputs a prime
number p, a public key PK, and a secret key SK.*

   – Sign(SK, id, $\mathbf{m}$). *This algorithm takes as inputs a secret key
SK, a file identifier* id $\in \{0,1\}^k$, *and a vector* $\mathbf{m} \in \mathbb{F}_p^N$ *and
supplies as output a signature σ.*

   – Combine(PK, id, $\{(\omega_i, \sigma_i)\}_{i=1}^n$). *This algorithm takes as
inputs a public key PK, a file identifier* id $\in \{0,1\}^k$, *and a
set* $\{\omega_i, \sigma_i\}_{i=1}^n$ *with $\omega_i \in \mathbb{F}_p$. It supplies as output a signature
σ. (The intuition is that if each σ_i is a valid signature of vector
$\mathbf{m}_i$, then σ is a signature of $\sum_{i=1}^n \omega_i \mathbf{m}_i$).*

   – Verify(PK, id, $\mathbf{y}, \sigma$). *This algorithm takes as inputs a public
key PK, an identifier* id $\in \{0,1\}^k$, *a vector* $\mathbf{y} \in \mathbb{F}_p^N$, *and a
signature σ. It supplies as output a Boolean 0 (reject) or 1
(accept).*

*It is required that for each triplet (p, PK, SK) output of
Setup($1^k, N$), the following properties are fulfilled:*

   – *For all* id *and all* $\mathbf{y} \in \mathbb{F}_p^N$, *if $\sigma \leftarrow$* Sign(SK, id, $\mathbf{y}$) *then*
Verify(PK, id, $\mathbf{y}, \sigma$) $= 1$.

   – *For all* id $\in \{0,1\}^k$ *and all triplet* $\{(\omega_i, \sigma_i, \mathbf{m}_i)\}_{i=1}^n$, *if the
following equality is fulfilled* Verify(PK, id, $\mathbf{m}_i, \sigma_i$) $= 1$ *for all i,
then*

$$\text{Verify}(PK, \text{id}, \sum_{i=1}^n \omega_i \mathbf{m}_i, \text{Combine}(PK, \text{id}, \{(\omega_i, \sigma_i)\}_{i=1}^n)) = 1.$$

Another definition of homomorphic signature, which is
more restrictive, has previously been suggested in [JOH 02].

5.4.2.2. *Examples of such schemes*

In [BON 09], the authors suggest two constructions that
fulfill the previous definition for a single source node.

These constructions are made from bilinear coupling on elliptical curves. The first scheme suggested, NCS_1, has an advantage that the signatures generated can be associated with individual vectors more than with an integer subspace. The second construction, NCS_2, supplies an instantiation of the scheme of [KRO 04], which is proven in the sense of the definition. We can consider the former construction as a more efficient version of the scheme suggested in [ZHA 07].

In [LI 10], the authors show that the construction suggested in [GEN 06], which is a signature scheme based on the identity, fulfills the previous definition and is hence a possible signature solution for network coding.

In both [AGR 10] and [BON 11], the authors extend the signature suggestions for network coding made in [BON 09] to the cases of multi-source and multi-file network coding through a generic construction. Furthermore, in [BON 11], the authors build signatures for network coding on binary fields by using Euclidean networks. These signatures cannot be built otherwise, that is by using, for instance, bilinear coupling or other conventional algebraic methods based on factorization or the discrete logarithm problem.

5.4.3. *Alternatives for integrity and authenticity*

The systems of network coding are vulnerable to pollution attacks in which a harmful node floods the network with polluted packets and prevents the recipient from properly decoding the received packets. A conventional method to prevent this attack is the method presented previously wherein we protect the integrity of the messages, validate the identity of the source, and warrant the non-rejection of information sent by the former. This implies the use of homomorphic signature schemes, homomorphic MAC, or homomorphic hash functions as described in sections 5.4.1 and 5.4.2. These solutions are efficient within the framework of network coding even if the signature schemes often have a high calculation cost, which cannot be necessarily supportable for small objects with constrained resources.

In this section, we will be interested in other existing solutions suggested in the literature to overcome the pollution attack problem. For this purpose, these methods rely on the use of checksums, polynomial distribution, and overlapping MAC to supply authentication solutions with a relatively small cost.

5.4.3.1. *Polynomial method*

In [OGG 08], the authors suggest a polynomial distribution method. They assume that there exist a source S, V receiving nodes, and a reliable authority in the network. The authority generates a secret matrix $\mathbf{A}$ of size $k \times (M + 1)$, which will represent the coefficients of $M + 1$ polynomials:

$$\mathbf{A} = \begin{pmatrix} a_{00} & a_{10} & \cdots & a_{M0} \\ a_{01} & a_{11} & \cdots & a_{M1} \\ \vdots & \vdots & \cdots & \vdots \\ a_{0k-1} & a_{1k-1} & \cdots & a_{Mk-1} \end{pmatrix},$$

The authority also generates V distinct values $r_1, r_2, \cdots, r_V \in \mathbb{F}_q$ that it distributes to each of the V receivers by making them public for all the nodes. Then, $M + 1$ secret polynomials are built by the source from matrix $\mathbf{A}$ and each receiver receives these $M + 1$ polynomials secretly. Each receiving node can hence calculate the evaluation of $M + 1$ polynomials:

$$(P_0(r_i), P_1(r_i), \cdots, P_M(r_i)) = (r_i^0, r_i^1, \cdots, r_i^{k-1}) \cdot \mathbf{A}.$$

When node S sends n messages $s_1, s_2, \cdots, s_n$, it calculates n polynomials $A_{s_i}(r)$ for message i, by using a public value r corresponding to a receiver:

$$A_{s_i}(r) = P_0(r) + s_i P_1(r) + s_i^q P_2(r) + \cdots + s_i^{q^{(M-1)}} P_M(r),$$

for i from 1 to n.

This authentication is to add to packet x_i with $x_i = (1, s_i, A_{s_i}(r))$, for $i = 1, \cdots, n$. By using a vector $v_i = (v_{i1}, v_{i2}, \cdots, v_{in})$, S encodes these packets:

$$y_i(x) = v_i \cdot \begin{pmatrix} x_1 \\ x_2 \\ \vdots \\ x_n \end{pmatrix} = \left(\sum_{j=1}^{n} v_{ij}, \sum_{j=1}^{n} v_{ij} s_j, \sum_{j=1}^{n} v_{ij} A_{s_i}(r) \right)$$

Thus, for each encoded packet received, receiving node R can check the authentication $\sum_{j=1}^{n} v_{ij} A_{s_i}(r)$ by calculating the polynomials itself.

In [OGG 08], the authors also show that this scheme is robust against a coalition of $k - 1$ harmful receivers among V receivers. This is because the following polynomial system

$$\mathbf{P} = \begin{pmatrix} P_0(x_1) & P_1(x_1) & \cdots & P_M(x_1) \\ P_0(x_2) & P_1(x_2) & \cdots & P_M(x_2) \\ \vdots & \vdots & \cdots & \vdots \\ P_0(x_k) & P_1(x_k) & \cdots & P_M(x_k) \end{pmatrix},$$

built from M receivers can be recovered only if a submatrix A' is built from T lines of secret matrix $\mathbf{A}$,

$$\mathbf{A} = \begin{pmatrix} a_{00} & a_{10} & \cdots & a_{T0} \\ a_{01} & a_{11} & \cdots & a_{T1} \\ \vdots & \vdots & \cdots & \vdots \\ a_{0k} & a_{1k} & \cdots & a_{Tk} \end{pmatrix}.$$

Directly, we see that T must be greater than $k-1$ to be able to ensure the reconstruction of **A** and hence the success of an attack by collusion.

5.4.3.2. *Method using checksums*

In [DON 09a], the authors suggest the use of a method based on checksums referred to as DART to supply a verification at low cost of received packets. This method uses only a simple pseudo-random function to generate the checksums at the level of the source node and to check that these sums are at the level of either the intermediate nodes or the destination. To avoid pollution attacks, the authors of [DON 09a] recommend using DART at the source node and the checking of the packets by each of the nodes of the path.

For each message that must be sent $G = [\overrightarrow{p_1}, \overrightarrow{p_2}, \cdots, \overrightarrow{p_n}]$, the source node generates a packet of checksum in the following way:

$$CHK_s(G) = H_s G$$

where $H_s = [u_{i,j}]$ is a random matrix of size $b \times n$ generated by the source by using a pseudo-random function f and a random seed s of k bits with $u_{i,j} = f_s(i||j)$. Then, the source includes $(CHK_s(G), s, t)$ to each packet with t a timestamp. Thus, the packets are disseminated to be identified. The checksums are periodically transferred to the following hops.

When a node receives a "fresh" checksum (at time $t - \Delta$) and a coded packet $(\overrightarrow{c}, \overrightarrow{e})$, with $\overrightarrow{c}$ the vector of coefficients of coded packet $\overrightarrow{e}$, it checks the packet by using matrix $CHK_s(G)$:

$$CHK_s(G)\overrightarrow{c} = (H_s G)\overrightarrow{c} = H_s(G\overrightarrow{c}) = H_s \overrightarrow{e}$$

In DART, when an intermediate node receives a coded packet, it must transfer the new packet after having checked the old packets. Thus, each transfer will be delayed because of the latency necessary to the verification. In the same paper [DON 09a], the authors suggest improving DART in a version named EDART. EDART allows the intermediate nodes to transfer new encoded packets without immediate verification. The packet is supplied with a decreasing hop counter and the verification will be carried out by the node receiving the zero counter.

If we compare DART with other schemes based on digital signatures like [CHA 06] and [YU 08], it is obvious that DART is much faster because it uses only an evaluation of a pseudo-random function f_s and vectorial multiplications. Thus, DART will be very efficient in constrained environments such as wireless sensor networks.

In DART, the intermediate nodes can only authenticate the source but cannot determine who has possibly polluted the packets. Hence, DART does not enable us to perform what is commonly referred to in cryptography as traitor tracing.

5.4.3.3. *Overlapping MAC*

In [YU 09], the authors suggest using "overlapped MAC" to secure network coding using an XOR operator against pollution attacks. For this purpose, they use pre-distribution of probabilistic keys, where each source will generate MACs for each of the messages to be sent. The intermediate nodes and the destination are thus able to authenticate MACs by using keys that they share with the source node. Unlike schemes using MACs, here, each source node generates several MACs for the same message and each MAC can authenticate only a part of the message with these parts overlapping.

In [YU 09], all the nodes are assumed to have been pre-distributed with a certain number of random secret keys using a method of key pre-distribution like the method described in [ESC 02].

Each node chooses a certain number of keys coming from a larger set. These keys are used to generate the MACs, whereas the intermediate nodes and the destination use the shared keys to verify these MACs.

In network coding using an XOR operator, a coded message is represented as $E = \alpha_1 M_1 \oplus \cdots \oplus \alpha_n M_n$, with $\alpha_i \in \{0, 1\}$ for $i = 1, \cdots, n$. Each message is divided into m codewords and expressed by a line vector $M_i = (m_{i,1}, \cdots, m_{i,m})$, where $m_{i,j}$ for $j = 1, \cdots, m$ represents the codewords.

To each message sent by the source, we attach t MACs and each MAC is generated from u codewords that are randomly chosen. A MAC for message M_i is calculated as follows: $MAC_{i,j} = \{id(k_{s,j}), r_j, h_{i,j}\}_{k_{s,j}}$, where $\{\cdot\}_{k_{s,j}}$ represents the MAC calculated with the help of key $k_{s,j}$, $h_{i,j}$ is the XOR of u codewords randomly chosen and r_j is a random number, the seed of a hash chain. All the MACs for M_i are attached to the coded message and each intermediate node and the destination can verify some of the MACs when they share keys with the source.

This MAC scheme warrants that the MACs can authenticate overlapping codewords, that is, instead of being identified by a single MAC, each codeword can be authenticated by $\frac{t \times u}{m}$ MACs. In some cases, this enables us to prevent compromised nodes from polluting the packets because the MACs generated by the intermediate nodes involve more keys than those generated by the compromised nodes. [YU 09] also supplies a mean path length of polluted packets before being detected.

5.5. Conclusion

As emphasized in this chapter, the creation of new information exchange methods is clearly combined to the creation of new dedicated attacks. It is hence necessary to develop new and specific methods to protect ourselves against these new attacks.

Network coding has opened new particularly interesting cryptographic ways that essentially consist of defining new homomorphic primitives to warrant confidentiality and integrity/authenticity. The cryptographic community has taken up the challenge by mainly suggesting new integrity and authenticity mechanisms that enable us to prevent pollution attacks in network coding. Similarly, even if this research axis is much more recent, some preliminary solutions have been suggested to warrant confidentiality through completely homomorphic encryption methods. It is clear that these solutions are less finalized than the previous solutions but show a promising future.

There remains one last thing to mention regarding the tsunami attacks. To our knowledge, only an article in [DON 09b] reports this particular problem without supplying all the solutions that enable us to prevent these attacks in the network. It is a study that is more devoted to potential threats and some analyses of the effects of tsunami attack. The previously cited article presents performance and energy-consumption degradations induced in a network subjected to a tsunami attack. Finding solutions to the tsunami attack problem hence seems to be a research path still open in the field of security for network coding.

5.6. Bibliography

[AGR 09] AGRAWAL S., BONEH D., "Homomorphic MACs: MAC-based integrity for network coding", *ACNS*, Springer, Paris-Rocquencourt, France, vol. 5536, pp. 292–305, 2–5 June 2009.

[AGR 10] AGRAWAL S., BONEH D., BOYEN X., FREEMAN D.M., "Preventing pollution attacks in multi-source network coding", *Public Key Cryptography*, Springer, Paris, France, vol. 6056, pp. 161–176, 26–28 May 2010.

[ANY 10] ANYA A., ZNAIDI W., FRABOULET A., GOURSAUD C., LAURADOUX C., MINIER M., "Energy friendly integrity for network coding in wireless sensor networks", IEEE, *International Conference on Network and System Security – NSS 2010*, Melbourne, Australia, pp. 1–8, 2010.

[BAU 05] BAUER F.L., "Vernam cipher", *Encyclopedia of Cryptography and Security*, Springer, 2005.

[BER 08] BERBAIN C., BILLET O., CANTEAUT A., COURTOIS N., GILBERT H., GOUBIN L., GOUGET A., GRANBOULAN L., LAURADOUX C., MINIER M., PORNIN T., SIBERT H., "Sosemanuk, a fast software-oriented stream cipher", *The eSTREAM Finalists*, Lecture Notes in Computer Science, Springer, pp. 98–118, 2008.

[BON 07] BONEH D., "A brief look at pairings based cryptography", *FOCS*, IEEE Computer Society, Rhode Island, USA, pp. 19–26, 20–23 October 2007.

[BON 09] BONEH D., FREEMAN D., KATZ J., WATERS B., "Signing a linear subspace: signature schemes for network coding", *Public Key Cryptography – PKC 2009*, Lecture Notes in Computer Science 5443, Springer, pp. 68–87, 2009.

[BON 11] BONEH D., FREEMAN D.M., "Linearly homomorphic signatures over binary fields and new tools for lattice-based signatures", *Public Key Cryptography*, Springer, Taormina, Italy, vol. 6571, pp. 1–16, 6–9 March 2011.

[CAN 99] CANETTI R., GARAY J.A., ITKIS G., MICCIANCIO D., NAOR M., PINKAS B., "Multicast security: A taxonomy and some efficient constructions", *INFOCOM*, IEEE, pp. 708–716, 1999.

[CAR 79] CARTER L., WEGMAN M.N., "Universal classes of hash functions", *Journal of Computer and System Sciences – JCSS*, vol. 18, no. 2, pp. 143–154, 1979.

[CHA 06] CHARLES D., JAIN K., LAUTER K., "Signatures for network coding", *40th Annual Conference on Information Sciences and Systems, 2006*, pp. 857–863, March 2006.

[DIJ 10] VAN DIJK M., GENTRY C., HALEVI S., VAIKUNTANATHAN V., "Fully homomorphic encryption over the integers", *EUROCRYPT*, Springer, vol. 6110, pp. 24–43, 30 May–3 June 2010.

[DON 09a] DONG J., CURTMOLA R., NITA-ROTARU C., "Practical defenses against pollution attacks in intra-flow network coding for wireless mesh networks", *ACM Conference on Wireless Network Security – WiSec '09*, ACM, pp. 111–122, 2009.

[DON 09b] DONG J., CURTMOLA R., NITA-ROTARU C., "Secure network coding for wireless mesh networks: threats, challenges, and directions", *Computer Communication*, vol. 32, no. 17, pp. 1790–1801, 2009.

[EKD 02] EKDAHL P., JOHANSSON T., "A new version of the stream cipher SNOW", *Selected Areas in Cryptography*, Springer, St. John's, Newfoundland, Canada, pp. 47–61, 15–16 August 2002.

[ESC 02] ESCHENAUER L., GLIGOR V., "A key-management scheme for distributed sensor networks", *Proceedings of the 9th ACM Conference on Computer and Communications Security*, ACM, pp. 41–47, 2002.

[FAN 09] FAN Y., JIANG Y., ZHU H., SHEN X., "An efficient privacy-preserving scheme against traffic analysis attacks in network coding", *INFOCOM*, IEEE, Brazil, pp. 2213–2221, 19–25 April 2009.

[FON 07] FONTAINE C., GALAND F., "A survey of homomorphic encryption for non-specialists", *EURASIP Journal on Information Security*, vol. 1, pp. 1–15, 2007.

[GEN 06] GENTRY C., RAMZAN Z., "Identity-based aggregate signatures", *Public Key Cryptography*, Springer, New York, USA, vol. 3958, pp. 257–273, 24–26 April 2006.

[GEN 09] GENTRY C., "Fully homomorphic encryption using ideal lattices", *STOC*, ACM, Bethesda, USA, pp. 169–178, 31 May–2 June 2009.

[GKA 05] GKANTSIDIS C., RODRIGUEZ P., "Network coding for large scale content distribution", *INFOCOM*, IEEE, pp. 2235–2245, 2005.

[GKA 06] GKANTSIDIS C., RODRIGUEZ P., "Cooperative security for network coding file distribution", *INFOCOM*, IEEE, 2006.

[GOL 82] GOLDWASSER S., MICALI S., "Probabilistic encryption and how to play mental poker keeping secret all partial information", *STOC*, ACM, pp. 365–377, 1982.

[JOH 02] JOHNSON R., MOLNAR D., SONG D.X., WAGNER D., "Homomorphic signature schemes", *CT-RSA*, pp. 244–262, 2002.

[KNU 73] KNUTH D., *The Art of Computer Programming. vol. 1, Fundamental Algorithms*, Addison-Wesley Pub. Co, 1973.

[KRA 94] KRAWCZYK H., "LFSR-based hashing and authentication", *Advances in Cryptology – CRYPTO '94*, Lecture Notes in Computer Science 839, Springer, pp. 129–139, 1994.

[KRO 04] KROHN M.N., FREEDMAN M.J., MAZIÈRES D., "On-the-fly verification of rateless erasure codes for efficient content distribution", *IEEE Symposium on Security and Privacy*, pp. 226–240, 2004.

[LI 10] LI Z., GONG G., "Data aggregation integrity based on homomorphic primitives in sensor networks", *ADHOC-NOW*, Springer, pp. 149–162, 2010.

[NAT 01] NATIONAL INSTITUTE OF STANDARDS AND TECHNOLOGY, FIPS PUB 197, Advanced Encryption Standard (AES), Federal Information Processing Standards Publication 197, U.S. Department of Commerce, November 2001.

[OGG 08] OGGIER F., FATHI H., "Multi-receiver authentication code for network coding", *46th Annual Allerton Conference on Communication, Control, and Computing, 2008*, IEEE, pp. 1225–1231, 2008.

[PAI 99] PAILLIER P., "Public-key cryptosystems based on composite degree residuosity classes", *EUROCRYPT*, Springer, pp. 223–238, 1999.

[SMA 10] SMART N.P., VERCAUTEREN F., "Fully homomorphic encryption with relatively small key and ciphertext sizes", *Public Key Cryptography – PKC 2010, 13th International Conference on Practice and Theory in Public Key Cryptography, Proceedings*, Springer, Paris, France, vol. 6056, pp. 420–443, 26–28 May 2010.

[STE 10] STEHLÉ D., STEINFELD R., "Faster fully homomorphic encryption", *ASIACRYPT*, Springer, pp. 377–394, 2010.

[YU 08] YU Z., WEI Y., RAMKUMAR B., GUAN Y., "An efficient signature-based scheme for securing network coding against pollution attacks", *INFOCOM 2008*, IEEE, Phoenix, USA, pp. 1409–1417, 13–18 April 2008.

[YU 09] YU Z., WEI Y., RAMKUMAR B., GUAN Y., "An efficient scheme for securing XOR network coding against pollution attacks", *INFOCOM 2009*, IEEE, pp. 406–414, 2009.

[ZHA 07] ZHAO F., KALKER T., MÉDARD M., HAN K.J., "Signatures for content distribution with network coding", *Proceedings of International Symposium on Information Theory (ISIT)*, IEEE, 2007.

[ZHA 10] ZHANG P., JIANG Y., LIN C., FAN Y., SHEN X., "P-Coding: Secure network coding against eavesdropping attacks", *INFOCOM*, IEEE, pp. 2249–2257, 2010.

Chapter 6

Random Network Coding and Matroids

In this chapter, we focus on different models for transmitting data by non-coherent network coding. Transmission is viewed as the communication of an invariant of the message that depends on the protocol being used. In particular, data communication is modeled as the transmission of flats of matroids. This generalization allows us to analyze and compare the performance of different protocols. As such, we will study the rate, delay and throughput of routing, random linear network coding (RLNC) and random affine network coding (RANC). This matroid-based model enables the non-coherent correction of errors. This is because errors in packages or messages are transformed into modifications of the transmitted flat. The problem is therefore reduced to the design of error-correcting codes on the flats. Using algebraic codes, notably Gabidulin codes, error correction can be carried out effectively. We will,

---

Chapter written by Maximilien GADOULEAU.

therefore, describe a high performance decoding algorithm for error-correcting codes for RANC.

6.1. Protocols for non-coherent communication

Let us first re-examine the three protocols for non-coherent communications. By non-coherent, we mean a mode of transmission in a network that does not take advantage of any knowledge of the network's characteristics. This is particularly desirable for dynamic networks such as wireless networks. The mode of transmission is therefore resistant to variations in the network's topology such as the appearance or disappearance of nodes or links. Similarly, intermediary nodes are meant to operate irrespective of the elements that they receive without accounting for the source, destination, or nature of the transmitted data.

The main concept of the model for non-coherent data transmission is described in the following. When transmitted over a network, the data are modified (this is obvious for network coding, but this is also the case for routing where the order of packets at the destination can be different from that at the source). Different protocols preserve different properties of the transmitted message: routing preserves the non-ordered set of packets, linear network coding preserves the subspace generated by the packets, etc. This is exactly the property preserved by the protocol that is effectively transmitted. We therefore see the communication of an invariant of the message; this invariant can be seen in general as a flat of a matroid.

6.1.1. Routing

Routing, also called store and forward (SAF), is the traditional means of transmitting data on a network. In this scenario, the intermediary nodes simply retransmit the received packets to their destination without combining them. On the other hand, due to the dynamic topology of the network or links of different delays, we will suppose that the packets are not received in the order in which they were sent by the source. Since the order of packets in the original message is lost during communication, SAF can be seen as the transmission without errors of the set of packets rather than the ordered sequence of packets [GAD 10a]. The traditional approach to counter the fact that the order of packets is lost is to add a header containing the position of the packet in the transmitted message to order them. For example, if the source wants to send three messages $\mathbf{m}_0, \mathbf{m}_1$ and $\mathbf{m}_2$ on $\mathrm{GF}(q)$ seen as the rows of a matrix $\mathbf{M}$, they transmit the following packets, known as the lifting of $\mathbf{M}$:

$$\begin{pmatrix} 0 & \mathbf{m}_0 \\ 1 & \mathbf{m}_1 \\ 2 & \mathbf{m}_2 \end{pmatrix}.$$

This is equivalent to the approach typically used in IP-type protocols. Each destination receives a matrix obtained by permutating rows of the matrix sent by the source. Decoding is straightforward.

6.1.2. *Random linear network coding*

RLNC is an innovative means of transmitting data in a non-coherent way [HO 06]. In this case, the packets are viewed as vectors of length n over a finite field $\mathrm{GF}(q)$. The intermediary nodes, therefore, choose at random a linear combination of the packets that they receive. While RLNC does not guarantee that all the information has been correctly transmitted in a minimum amount of time, algebraic arguments show that for fairly large fields, the probability of receiving the message in its entirety in minimal time is close to 1. On the other hand, as the complexity of the combination and decoding increases with the size of field, $GF(q)$, where $q = 2^8$ or $q = 2^{16}$, are generally suggested for possible application.

The source should encode its messages in linearly independent vectors. Contrary to the typically proposed approach, we will restrict ourselves to non-zero vectors whose first non-zero coordinate is equal to 1. This convention will be explained later. To encode its messages, the source can use linear lifting [SIL 08], defined as adding a header formed by the identity matrix:

$$\begin{pmatrix} 1 & 0 & 0 & m_0 \\ 0 & 1 & 0 & m_1 \\ 0 & 0 & 1 & m_2 \end{pmatrix}$$

The messages thus encoded in k linearly independent packets in $GF(q)^n$ are transmitted on the network. Since the linear combinations operated by the intermediary nodes do not modify the subspace generated by the transmitted matrix, RLNC is seen as the transmission of a linear subspace of dimension k of $GF(q)^n$ [KOE 08]. More precisely, the source sends the matrix $(\mathbf{I}_k|\mathbf{M})$ and each destination will receive a matrix $(\mathbf{L}|\mathbf{LM})$, where $\mathbf{L}$ is chosen randomly (and each destination receives a different matrix $\mathbf{L}$) [SIL 10b]. If the matrix $\mathbf{L}$ is chosen uniformly among all the matrices of $GF(q)^{k \times k}$, then the probability that $\mathbf{L}$ is invertible is greater than

$$K_q = \prod_{i=1}^{\infty}(1 - q^{-i}),$$

which tends to 1 when q tends to infinity. Decoding can be done easily by finding the inverse of $\mathbf{L}$.

6.1.3. *Random affine network coding*

With RANC, the packets are seen as points in an affine space and the intermediary nodes combine the packets by affine combinations. An affine combination of the points $v_0, v_1, \ldots, v_{k-1} \in GF(q)^n$ is any linear combination of the form

$$\sum_{i=0}^{k-1} a_i v_i, \qquad \text{where} \quad \sum_{i=0}^{k-1} a_i = 1.$$

In other words, an affine combination corresponds to identifying the centroid of the points v_i with weights a_i.

Let us now examine the blueprint to implement RANC. Firstly, encoding messages (seen as the rows m_i of a matrix $M \in GF(q)^{k \times (n-k+1)}$) in affinely independent points can be simply carried out by adding the header $I'_k = (0|I_{k-1})^T$ to obtain $(I'_k|M) \in GF(q)^{k \times n}$. Note that the header is shorter than that used for RLNC by a symbol, which illustrates the advantage offered by RANC. We will call this encoding the affine lifting of M. For example, for the three messages already studied for SAF and RLNC, we obtain

$$\begin{pmatrix} 0 & 0 & m_0 \\ 1 & 0 & m_1 \\ 0 & 1 & m_2 \end{pmatrix}.$$

Secondly, as affine combinations are particular linear combinations, the complexity of RANC for intermediary nodes is of the same order as that for RLNC. We should note here that there is no restriction on the affine combinations operated on the intermediary nodes: every node (including the source) can send the origin, for example.

Thirdly, we will describe the decoding algorithm at the destination, therefore showing that this does not increase complexity. Let us suppose that the destination receives k

affinely independent points $\mathbf{v}_0, \mathbf{v}_1, \ldots, \mathbf{v}_{k-1}$, then the first k columns of $(\mathbf{1}|\mathbf{V}) \in \mathrm{GF}(q)^{k \times (n+1)}$ are linearly independent, where $\mathbf{1}$ is the all-ones column vector. Gaussian elimination on $(\mathbf{1}|\mathbf{V})$ therefore produces $(\mathbf{I}_k|\mathbf{M}')$, where $\mathbf{M}' = (\mathbf{m}_0^T, (\mathbf{m}_1 - \mathbf{m}_0)^T, \ldots, (\mathbf{m}_{k-1} - \mathbf{m}_0)^T)^T$. Decoding is concluded by adding $\mathbf{m}_0$ to all the other rows of $\mathbf{M}'$. The complexity of the algorithm is therefore given by inverting a square matrix of order k, which is similar to the complexity of RLNC. Finally, note that Gaussian elimination can be modified to obtain the matrix $(\mathbf{1}|\mathbf{I}'_k|\mathbf{M})$ directly.

6.1.4. *Example and comparison*

We illustrate the differences among SAF, RLNC, and RANC with the butterfly network represented in Figure 6.1, where the source S wants to transmit two messages $\mathbf{m}$ and $\mathbf{n}$ over $\mathrm{GF}(q)$ to the destinations D and E.

If SAF is used, the source encodes the messages and orders them by adding the following headers: $\mathbf{r} = (0|\mathbf{m})$ and $\mathbf{s} = (1|\mathbf{n})$. The intermediary nodes can only select a message from $\mathbf{r}$ and $\mathbf{s}$. In this case, it is obvious that a single destination receives both messages while the other receives only one. As such, the probability of success is zero.

Let us now suppose that RLNC is used. The source encodes the messages in linearly independent vectors whose first non-zero coordinate is equal to 1 by adding the following headers: $\mathbf{u} = (10|\mathbf{m})$ and $\mathbf{v} = (01|\mathbf{n})$. The only linear combination of a vector is simply the same vector. All the linear combinations of $\mathbf{u}$ and $\mathbf{v}$ are either $\mathbf{v}$ or take the form $\mathbf{u} + a\mathbf{v}$, where $a \in \mathrm{GF}(q)$. There are therefore $q + 1$ possible combinations and $q - 1$ guarantee decoding at the destinations (if $a \neq 0$), which gives a probability of success of $1 - \frac{2}{q+1}$, tending to 1 for large values of q.

Lastly, let us suppose that RANC is used. The source encodes these messages in affinely independent points by adding the following headers: $\mathbf{x} = (0|\mathbf{m})$ and $\mathbf{y} = (1|\mathbf{n})$. The only linear combination of a point is this same point; all the affine combinations of $\mathbf{x}$ and $\mathbf{y}$ take the form $b\mathbf{x} + (1 - b)\mathbf{y}$, where $b \in \mathrm{GF}(q)$. There are therefore q possible combinations and $q - 2$ guarantee decoding at the destinations (if $b \notin \{0, 1\}$), which gives a probability of success of $1 - \frac{2}{q}$. This probability is zero for the binary field because in this very specific case, RANC is reduced to SAF. On the other hand, for large q, it tends to 1 almost as quickly as for RLNC. Lastly, note that decoding messages at the destination D, which receives the points $(0|\mathbf{m})$ and $(1 - b|b\mathbf{m} + (1 - b)\mathbf{n})$, is elementary (similarly for E): D constructs the matrix

$$\left(\begin{array}{cc|c} 1 & 0 & \mathbf{m} \\ 1 & 1 - b & b\mathbf{m} + (1 - b)\mathbf{n} \end{array} \right),$$

which, after modified Gaussian elimination, becomes

$$\left(\begin{array}{cc|c} 1 & 0 & \mathbf{m} \\ 1 & 1 & \mathbf{n} \end{array} \right).$$

6.2. Transmission model based on flats of matroid

6.2.1. *Matroids*

Let us review the definition and essential properties of matroids and their flats. While the concepts defined in the following originate from matroid theory, they can all be seen as generalizations of well-known concepts in linear algebra. For a detailed overview of matroid theory, see [OXL 06].

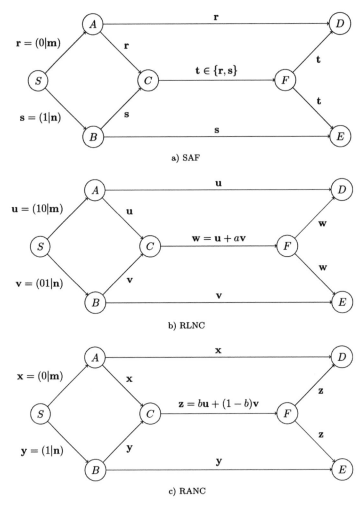

Figure 6.1. *Transmitting data on the butterfly network with SAF, RLNC, or RANC*

For a finite set E, we denote the set of subsets of E with cardinality $0 \leq i \leq |E|$ by $\mathcal{P}(E, i)$, and its power set by $\mathcal{P}(E) = \bigcup_{i=0}^{|E|} \mathcal{P}(E, i)$. A matroid is a pair $\mathcal{M} = (E, \mathcal{I})$, where E and $\mathcal{I} \subseteq \mathcal{P}(E)$ are, respectively, called the ground set and the independent sets of $\mathcal{M}$. The independent sets are generalizations of linearly independent vectors and satisfy the three following axioms:

- $\emptyset \in \mathcal{I}$;
- if $A \in \mathcal{I}$ and $B \subset A$, then $B \in \mathcal{I}$;
- if $I_1, I_2 \in \mathcal{I}$ with $|I_1| > |I_2|$, then there is $e \in I_1 \backslash I_2$ such that $I_2 \cup \{e\} \in \mathcal{I}$.

The third axiom, called independence augmentation axiom, is crucial because it guarantees that every family of independent elements can be extended to form a basis (a maximal family of independent elements). Clearly, all the bases have the same cardinality.

With each matroid there is an associated rank function $\mathrm{rk}(A)$ for every $A \subseteq E$, defined as the maximum number of independent elements in A. For all subsets $A, B \subseteq E$, we have the submodular inequality

$$\mathrm{rk}(A \cup B) + \mathrm{rk}(A \cap B) \leq \mathrm{rk}(A) + \mathrm{rk}(B).$$

The rank of a matroid is simply the rank of its ground set and is equal to the cardinality of a basis. The closure $\mathrm{cl}(A)$ of a subset A of elements is defined as the maximal subset of E containing A such that $\mathrm{rk}(\mathrm{cl}(A)) = \mathrm{rk}(A)$. The closure is the generalization of the span of a set of vectors and the rank is the generalization of the dimension of this span.

A flat is a set equal to its closure and therefore the generalization of a linear subspace. In particular, a flat with rank $r - 1$ in a matroid with rank r is called a hyperplane. By extension, we refer to every family of k independent elements in a flat with rank k as a basis of this flat. The set of flats of a matroid, written as $\mathcal{F}$, is closed by an intersection. We also denote the set of flats with rank k by $\mathcal{F}_k$ and its cardinality by N_k for every $0 \leq k \leq r$. Furthermore, the set of flats ordered by inclusion forms a lattice where the meeting of two flats is their intersection and their joining is the closure of their union.

A matroid can contain loops and parallel elements. A loop l is an element that does not belong to any independent set: $\{l\} \notin \mathcal{I}$; equally, l belongs to all the flats. Elements are parallel if they are pairwise dependent: $\{e_i, e_j\} \notin \mathcal{I}$ for $i \neq j$; they therefore belong to a set of rank 1. Loop and parallel elements are therefore generalizations of the all-zero vector and collinear vectors, respectively. A matroid is said to be simple if it contains neither loops nor parallel elements. For every matroid $\mathcal{M}$, the simple matroid obtained after removing all the loops and having kept a single element from each family of parallel elements from $\mathcal{M}$ has the same lattice of flats as $\mathcal{M}$. For every simple matroid, we have $\mathcal{F}_0 = \{\emptyset\}$, $N_0 = 1$, and $\mathcal{F}_1 = \mathcal{P}(E, 1)$, $N_1 = |E|$. In the absence of any other indication, all the matroids considered here are simple.

All the flats of the same rank do not necessarily have the same cardinality and the same number of bases. Nevertheless, matroids where all the flats of the same rank have equal cardinalities do exist and are called perfect matroid designs [DEZ 92]. Due to their specific structure, few classes of perfect matroid designs are known to date; however, we are going to examine only the perfect matroid designs. We will therefore denote the cardinality or every flat with rank k by C_k, where $C_0 = 0$ and $C_1 = 1$ for every simple perfect matroid design.

6.2.2. Model and comments

In this section, we will re-examine the communication model for transmitting non-coherent and errorless data on a network introduced in [GAD 11]. Let us imagine that a source wants to transmit a message M in the alphabet $[A] = \{0, 1, \ldots, A - 1\}$ on a network toward a set of destinations. $\mathcal{M} = (E, \mathcal{I})$ is a matroid with a set $\mathcal{F}_k$ of flats with rank k, and let us suppose that the source and each destination have an injection G from $[A]$ to $\mathcal{F}_k$. Without loss of generality, we will suppose that an element of the flat corresponds to one transmitted packet and that these packets have a length of n over $\mathrm{GF}(q)$.

The errorless transmission of data consists of three stages:

1) The source encodes the original message M into a flat $f = G(M) \in \mathcal{F}_k$ and transmits a flow of elements of f containing a basis of f.

2) Each intermediary node combines the received elements by selecting and retransmitting elements from their closure.

3) Each destination expects to receive a basis of f; it recovers the original message by identifying $M = G^{-1}(f)$.

The structure of the matroid ensures that the rank function is well behaved. This is because a flat with rank k can be described only by not more than nor fewer than k independent elements. The independence augmentation axiom guarantees that every set with fewer than k independent elements can be extended into a flat basis.

A non-simple matroid contains loops and parallel elements. By definition, a loop belongs to every flat and therefore does not carry any information on the transmitted flat. In addition, two parallel elements belong to the same flat and are combined in the same way; they therefore carry the same information. As a result, loops and parallel elements are unnecessary and the hypothesis of only considering simple matroids does not lose any generality.

While flats of any rank can be sent, the two following reasons justify the hypothesis of only sending flats of the same rank k. Firstly, this guarantees that no transmitted flat is completely contained in another, rendering decoding non-ambiguous. In addition, each destination expects the same number of independent elements before beginning decoding, which simplifies the decoding procedure.

In terms of practical applications, all the intermediary nodes must have an effective algorithm for combining the elements that, on the other hand, does not guarantee generating a new flat basis. This operation can be seen as a form of random sampling on the flat's elements. Furthermore, the destinations need to have an effective algorithm to find the original message from every flat basis. For a generic matroid, effective algorithms may not exist; nevertheless, we will only consider the matroids where the elements can be combined and decoded quickly.

6.2.3. *Matroids for SAF, RLNC, and RANC*

Let us now reconsider three important classes of matroids (all are simple perfect matroid designs) that we associate with the protocols examined in section 6.1.

Firstly, the uniform matroid on r elements, normally denoted by $U_{r,r}$, has $[r] = \{0, 1, \ldots, r-1\}$ as ground set and all the subsets of $[r]$ are independent. Clearly, this matroid has rank r and each subset of $[r]$ is a flat. As such, $N_k = \binom{r}{k}$ and $C_k = k$ for every $1 \leq k \leq r$. It is evident that SAF corresponds to the transmission of flats for the matroid U_{q^n,q^n}. To use notations reflecting the protocol as well as the alphabet and the length of packets, we denote U_{q^n,q^n} by $\mathcal{S}(q,n)$ or simply $\mathcal{S}$ if there is no ambiguity.

Secondly, the projective geometry $PG(r-1,q)$ has the non-zero vectors of $\mathrm{GF}(q)^r$, whose first non-zero coordinate is equal to 1 as elements, where linear independence is used. This matroid has rank r and its flats are in bijection with the linear subspaces of $GF(q)^r$. As such, the number of flats with rank k is the Gaussian binomial

$$N_k = \begin{bmatrix} r \\ k \end{bmatrix} = \prod_{i=0}^{k-1} \frac{q^{r-i} - 1}{q^{k-i} - 1} \sim q^{k(r-k)}.$$

More precisely, it satisfies $q^{k(r-k)} \leq \begin{bmatrix} r \\ k \end{bmatrix} < K_q^{-1} q^{k(r-k)}$ for every $0 \leq k \leq r$. Furthermore, we have $C_k = (q^k - 1)/(q - 1)$ for every $1 \leq k \leq r$.

As previously indicated, our model differs from the purely random combinations generally proposed for RLNC. A linear combination can produce the all-zero vector or collinear vectors, that is a loop or parallel elements. On the other hand, our model considers the simple matroid associated with RLNC, which is clearly $PG(n-1,q)$ and which we will denote by $\mathcal{L}(q, n)$ or simply $\mathcal{L}$.

Thirdly, removing a hyperplane from $PG(r-1,q)$ produces the affine geometry $AG(r-1,q)$. This matroid also has rank r and its flats are the affine subspaces of $GF(q)^{r-1}$; there are $N_k = q^{r-k} \begin{bmatrix} r-1 \\ k-1 \end{bmatrix} \sim q^{k(r-k)}$ flats with rank k, which contain $C_k = q^{k-1}$ points for every $1 \leq k \leq r$ [OXL 06, section 6.2]. Every affine subspace of rank k can be represented by a linear subspace with dimension $k - 1$ translated by a point belonging to a complimentary subspace. By definition, $AG(r-1,q)$ is a submatroid of $PG(r-1,q)$, and can be seen as a matroid on the points of $GF(q)^{r-1}$, where two points $\mathbf{u}, \mathbf{v}$ are affinely independent if and only if the vectors $(1, \mathbf{u}), (1, \mathbf{v}) \in GF(q)^r$ are linearly independent.

The set of all the possible affine combinations of a collection of points, called the affine hull, forms an affine subspace. A collection of points $\mathbf{v}_0, \mathbf{v}_1, \ldots, \mathbf{v}_{k-1} \in GF(q)^n$ is said to be affinely independent if $\sum_{i=0}^{k-1} b_i \mathbf{v}_i \neq 0$ for all b_i that are not all zero and that satisfy $\sum_{i=0}^{k-1} b_i = 0$. By definition, the rank of a set of points is given by the number of affinely independent points and is equal to the rank and their affine hull. For every $\mathbf{v}_0, \mathbf{v}_1, \ldots, \mathbf{v}_{k-1} \in \mathrm{GF}(q)^n$, we have $\mathrm{rk}(\mathbf{v}_0, \mathbf{v}_1, \ldots, \mathbf{v}_{k-1}) = \mathrm{rank}(\mathbf{1}|\mathbf{V})$, where $\mathbf{V} = (\mathbf{v}_0^T, \mathbf{v}_1^T, \ldots, \mathbf{v}_{k-1}^T)^T$ and rank denotes the number of linearly independent rows in a matrix.

The matroid associated with RANC is therefore the affine geometry $AG(n, q)$, which we will denote by $\mathcal{A}(q, n)$ or simply $\mathcal{A}$ if there is no ambiguity regarding the parameters' values. We know that the simple matroid associated with RLNC is the projective geometry of rank n, where the alphabet has only $\begin{bmatrix} n \\ 1 \end{bmatrix} \sim q^{n-1}$ elements. This involves a loss in terms of efficiency because the elements are not encoded in an optimal way in packets with length n. For the same reason, every linear subspace consists of $\begin{bmatrix} k \\ 1 \end{bmatrix} \sim q^{k-1}$ elements that when compared with q^k possible linear combinations induce a loss of combinations. These problems are the immediate consequence of the existence of a loop (the all-zero vector) and of parallel elements (collinear vectors). Contrary to RLNC, the matroid associated with RANC has rank $n + 1$ and q^n elements. By its construction, $\mathcal{A}(q, n) = AG(n, q)$ is a submatroid of $\mathcal{L}(q, n+1) = PG(n, q)$. We will therefore demonstrate that $\mathcal{A}(q, n)$ behaves as $\mathcal{L}(q, n + 1)$, therefore allowing us to virtually work on packets with a length of $n + 1$ instead of n.

6.3. Parameters for errorless communication

The model presented in section 6.2 is very general and provides a unified approach for distinct problems such as SAF, RLNC, and RANC. It may seem strange to associate routing with a transmission model adapted to network coding. Nevertheless, note that routing is only one specific case of network coding where the single combination authorized for each intermediary node is the selection of a packet. The model allows us to concentrate on the combinatorial properties of network protocols in terms of encoding and combinations of data. In addition, by linking a matroid to a protocol, we have a new means of comparing the performance of different protocols.

In this section, we will define, identify, and compare some performance parameters for different matroids corresponding to a comparison of the performances of different network protocols. To carry out this study, we will focus on the following hypothesis about the model described in section 6.2. Since the model is non-coherent, the network's topology and the statistic dependence between packets due to the order of combinations are transparent at the message level. As such, we assume that each destination receives independently and uniformly chosen elements from all the elements of the transmitted flat. While a destination does not necessarily receive the entire message immediately, it continues to receive elements of the flat and therefore will be almost surely able to reconstruct the entire flat. Communication remains "errorless" because no other flat of the same rank can be reconstructed by a destination and as such only the message sent by the source can be decoded. This hypothesis can be seen as a generalization of the multiplicative channel matrix proposed for RLNC in [SIL 10b]. Furthermore, this hypothesis offers a good insight into the influence of the matroid's parameters on the performance of the corresponding protocol. The results of this section are taken from [GAD 11].

6.3.1. *Rate, delay and throughput*

The data rate of the communication is given by the ratio between the amount of information decoded and the amount of data required to transmit a flat: $\frac{\log_q A}{nk} = R_{\text{code}}(A, \mathcal{M})R_{\mathcal{M}}$, where $R_{\text{code}}(A, \mathcal{M}) = \frac{\log_q A}{\log_q N_k}$ can be seen as the code rate of the code formed by all the possible flats transmitted and where the rate is defined by (where we omit dependence in n and k)

$$R_{\mathcal{M}} = \frac{\log_q N_k}{nk}. \qquad [6.1]$$

Note that $R_{\text{code}}(A, \mathcal{M})$ depends only on the encoding of the message into a flat and not on the matroid (we only want $N_k \geq A$). We will therefore focus only on the rate $R_{\mathcal{M}}$, which indicates the efficiency with which a flat with rank k is encoded into k packets. We can break down the rate into

$$R_{\mathcal{M}} = \frac{\log_q N_k}{k \log_q |E|} \cdot \frac{\log_q |E|}{n},$$

where the first ratio is an intrinsic property of the matroid whereas the second ratio indicates the effectiveness of packaging an element of the matroid. Note that the rate is entirely determined by the lattice of flats in $\mathcal{M}$, and does not depend on the cardinality of the flats. Proposition 6.1 shows the rates for SAF, RLNC, and RANC.

PROPOSITION 6.1. (Rate).– The rates for SAF, RLNC, and RANC are, respectively, given by

$$R_{\mathcal{S}} = \frac{\log_q \binom{q^n}{k}}{nk} \sim 1 - \frac{\log_q k}{n},$$

$$R_{\mathcal{L}} = \frac{\log_q \begin{bmatrix} n \\ k \end{bmatrix}}{nk} \sim 1 - \frac{k}{n},$$

$$R_{\mathcal{A}} = \frac{n - k + 1 + \log_q \begin{bmatrix} n \\ k-1 \end{bmatrix}}{nk} \sim 1 - \frac{k-1}{n}.$$

RANC offers an improvement in terms of rate of around one symbol per packet in relation to RLNC due to the increase in the number of flats from approximately $q^{k(n-k)}$ to around $q^{k(n-k+1)}$. This gain is because of the fact that RLNC only considers approximately q^{-1} from all the possible packets with length n, while RANC considers all the possible packets.

According to the previous hypothesis, the packets (elements of the matroid) arrive at their destinations randomly. Therefore, the number of elements to be received to obtain k independent elements, called the transmission delay, is a random variable. The minimum delay is exactly k, while the maximum delay is unbounded. We will therefore define the average delay of a transmission $D_{\mathcal{M}}$ as the expectation of the number of elements received to obtain k independent elements. Clearly, $D_{\mathcal{M}} \geq k$ for every matroid $\mathcal{M}$. By generalizing the approach typically used to resolve the coupon collector, we obtain

$$D_{\mathcal{M}} = \sum_{i=0}^{k-1} \frac{1}{1 - \frac{C_i}{C_k}} = k + \sum_{i=0}^{k-1} \frac{1}{\frac{C_k}{C_i} - 1}. \qquad [6.2]$$

We will now determine the value of the average delay for SAF, RLNC and RANC.

PROPOSITION 6.2. (Average delay).– The average delays for SAF, RLNC, and RANC are given by

$$D_{\mathcal{S}} = k \sum_{i=1}^{k} i^{-1} \sim k(\ln k + \gamma),$$

$$D_{\mathcal{L}} = k + \sum_{j=1}^{k-1} \frac{1 - q^{j-k}}{q^j - 1} \sim k,$$

$$D_{\mathcal{A}} = k + \sum_{j=1}^{k-1} \frac{1}{q^j - 1} \sim k,$$

where $\gamma \approx 0.577$ is the Euler constant.

Proposition 6.2 indicates that for RLNC and RANC, the average number of packets required to completely decode the subspace tends toward the rank of the subspace when q tends to infinity. The delay of RANC is very close to that of RLNC because, according to [6.2], the average delay is determined by the cardinality of flats that are similar for the two protocols.

We will now define the throughput of a matroid as the ratio between the amount of information transmitted and the amount of data received on average by each destination. In other words, it measures the proportion of useful information received by each destination. By definition, the throughput is given by

$$T_{\mathcal{M}} = \frac{\log_q N_k}{n D_{\mathcal{M}}} = k \frac{R_{\mathcal{M}}}{D_{\mathcal{M}}}. \qquad [6.3]$$

This provides an indication of the desirable properties for a matroid for communications on a network. By [6.3], a matroid should maintain a low average delay while attempting to maximize its rate. By [6.2], minimizing the average delay involves minimizing the ratio $\frac{C_i}{C_k}$; in addition, by [6.1] the rate increases with the number of flats N_k. A matroid should therefore have a large number of flats whose cardinals increase rapidly with their rank.

By combining the previous results, the throughputs for SAF, RLNC, and RANC are approximately

$$T_{\mathcal{S}} \sim \frac{1}{\ln k} - \frac{1}{n \ln q}, \quad T_{\mathcal{L}} \sim 1 - \frac{k}{n}, \quad T_{\mathcal{A}} \sim 1 - \frac{k-1}{n}.$$

The throughputs for RLNC and RANC are therefore higher for small values of k, but decrease linearly with k. On the other hand, the throughput for SAF is more appropriate for messages carrying a large number of packets. The improvements in rate and similar delay between RANC and RLNC produce a gain in terms of throughput of a symbol by packet.

6.3.2. *Number of independent elements received*

We now study throughput in further detail by considering the probability $P_{\mathcal{M}}(r, l)$ of obtaining l independent elements once r elements have been received. A particularly important example is the probability $P_\mathcal{M}(k, k)$ of receiving all the independent elements necessary for reconstructing the flat with minimal delay. In general, the formula $P_\mathcal{M}(r, l)$ is complex; however, a direct approach allows us to obtain relatively simple formulas for the probability of independence for SAF, RLNC, and RANC.

PROPOSITION 6.3. Probability of independence.– For every $l \geq 1$

$$
P_\mathcal{S}(r, l) = k^{-r} \binom{k}{l} \sum_{j=0}^{l} (-1)^{l-j} \binom{l}{j} j^r,
$$

$$
P_\mathcal{L}(r, l) = (q^k - 1)^{-r} \begin{bmatrix} k \\ l \end{bmatrix} \sum_{s=0}^{r-l} (-1)^s \binom{r}{s} \prod_{i=0}^{l-1} (q^{r-s} - q^i),
$$

$$
P_\mathcal{A}(r, l) = q^{-(k-1)(r-1)} \begin{bmatrix} k-1 \\ l-1 \end{bmatrix} \prod_{i=0}^{l-2} (q^{r-1} - q^i);
$$

in particular,

$$
P_\mathcal{S}(k, k) = \frac{k!}{k^k} \sim \sqrt{2\pi k} \, e^{-k},
$$

$$
P_\mathcal{L}(k, k) = \prod_{i=1}^{k-1} \left(1 - \frac{q^i - 1}{q^k - 1} \right) \sim 1,
$$

$$
P_\mathcal{A}(k, k) = \prod_{i=1}^{k-1} (1 - q^i) \sim 1.
$$

In fact, $P_\mathcal{L}(k, k)$ and $P_\mathcal{A}(k, k)$ are very close and both are greater than K_q. It is therefore almost certain that the first k received packets will be informative if RLNC or RANC is used. In contrast, it is almost impossible to receive a complete message with minimum delay for SAF.

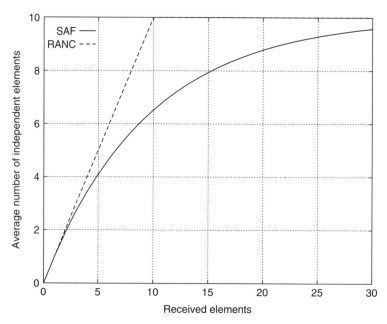

Figure 6.2. *Average number of independent elements according to the number of elements received for $k = 10$ transmitted elements*

We will now study the expectation $E_{\mathcal{M}}(r)$, that is the average number of independent elements among r elements received. Clearly, $E_{\mathcal{M}}(r) \leq \min\{r, k\}$ for every r. Proposition 6.4 determines or evaluates the expectation for SAF, RLNC, and RANC.

PROPOSITION 6.4. (Average number of independent elements).– For every k and r

$$E_{\mathcal{S}}(r) = k\left[1 - \left(1 - \frac{1}{k}\right)^{r}\right] \sim k(1 - e^{-\frac{r}{k}}). \qquad [6.4]$$

For RLNC and RANC, $E_{\mathcal{L}}(r)$ and $E_{\mathcal{A}}(r)$ are both greater than $K_q \min\{r, k\}$.

In particular, for $r = k$, [6.4] indicates that only around $1 - e^{-1} \approx 0.632$ of the first k elements received are independent on average when SAF is used. However, for RLNC and RANC, the average number of independent elements tends toward the optimal with the size of the field being used. This is illustrated in Figure 6.2 for $q = 2^8$, $k = 10$, and $1 \leq r \leq 30$. Note that $P_{\mathcal{M}}(r, l)$ does not depend on n for $\mathcal{M} \in \{\mathcal{S}, \mathcal{L}, \mathcal{A}\}$. For SAF, the inverse exponential form is clearly illustrated while RANC is almost optimal for practical values of q. Proposition 6.2 indicates that in this case the average throughput $D_{\mathcal{S}} \approx 29.3$ for SAF and $D_{\mathcal{A}} \approx 10.004$ for RANC.

For the three protocols studied, the characteristics of the matroids seen in section 6.2.3 and the aforementioned performance parameters are summarized in Table 6.1.

Protocol	SAF	RLNC	RANC
Matroid	U_{q^n, q^n}	$PG(n-1, q)$	$AG(n, q)$
Matroid rank	q^n	n	$n+1$
Combinations	Selections	Linear	Affine
Flats	Subsets	Linear subspaces	Affine subspaces
Number of flats with k	$\binom{q^n}{k}$	$\begin{bmatrix} n \\ k \end{bmatrix}$	$q^{n-k+1} \begin{bmatrix} n \\ k-1 \end{bmatrix}$
Cardinality of flats with rank k	k	$\frac{q^k - 1}{q - 1}$	q^{k-1}
Rate	$\sim 1 - \frac{\log_q k}{n}$	$\sim 1 - \frac{k}{n}$	$\sim 1 - \frac{k-1}{n}$
Average delay	$\sim k \ln k$	$\sim k$	$\sim k$
Throughput	$\sim \frac{1}{\ln k} - \frac{1}{n \ln q}$	$\sim 1 - \frac{k}{n}$	$\sim 1 - \frac{k-1}{n}$
Probability of k independent	$\sim \sqrt{2\pi k}\, e^{-k}$	~ 1	~ 1
Independent elements	$\sim k(1 - e^{-\frac{r}{k}})$	$\sim \min\{r, k\}$	$\sim \min\{r, k\}$

Table 6.1. *Summary of the parameters for SAF, RLNC and RANC*

6.4. Error-correcting codes for matroids

The matroid-based model also allows us to consider techniques for correcting errors when the network is not perfect. Several techniques have been proposed for correcting errors with SAF and RLNC (see [YEU 06, CAI 06] for coherent error correction) but here we will focus on the operator channel presented by [KOE 08] for RLNC and generalized for matroids in [GAD 11]. By considering every alteration in the message as a modification of the flat, the matroid codes can as such indifferently correct errors in each packet or in the entire message.

6.4.1. *Operator channel and lattice distance*

The model presented in section 6.2 does not take into account the possible alterations created by the message during its transmission. These alterations, due to lost, injected or erroneous packets, modify the transmitted flat. Formally, a flat $f \in \mathcal{F}_k$ can be modified in two ways: a deletion changes f into a sub-flat of rank $k - 1$, while an insertion changes f into a supra-flat containing f with rank $k + 1$. These changes correspond to moving up (or down) one step on the lattice of flats. Any flat f can be changed into any other g via a sequence of insertions and deletions. Proposition 6.5 proves that the shortest path for modifying a flat is to practice all the insertions first and then all the deletions subsequently.

PROPOSITION 6.5. Lattice distance.– For every pair of flats f, g in a matroid $\mathcal{M}$, the shortest path consists of moving up the lattice from f until $\mathrm{cl}(f \cup g)$ and then going down toward g. As such, the distance between f and g is given by

$$2\mathrm{rk}(f \cup g) - \mathrm{rk}(f) - \mathrm{rk}(g).$$

We will therefore model the transmission of data across a network as an operator channel, where the source transmits a flat $f \in \mathcal{F}$ and each destination receives another flat $g \in \mathcal{F}$, obtained after $\mathrm{rk}(f \cup g) - \mathrm{rk}(f)$ insertions and $\mathrm{rk}(f \cup g) - \mathrm{rk}(g)$

deletions. We will define the lattice distance between f and g by

$$d_{\mathcal{M}}(f, g) = 2\mathrm{rk}(f \cup g) - \mathrm{rk}(f) - \mathrm{rk}(g) \qquad [6.5]$$

$$\leq \mathrm{rk}(f) + \mathrm{rk}(g) - 2\mathrm{rk}(f \cap g), \qquad [6.6]$$

where [6.6] originates from the submodular inequality.

The term on the right-hand side of [6.6] corresponds to the number of deletions and insertions required to go from f to g if all the deletions are carried out first; in general, this term is not equal to the lattice distance between f and g [OXL 06, Proposition 6.5]. For example, two parallel hyperplanes f and g in $\mathcal{A}(q, n)$ have a distance of 2, while [6.6] gives $2n$. Furthermore, by considering f, g, and $\mathrm{GF}(q)^n$, we can easily demonstrate that the term on the right-hand side of [6.6] violates the triangular inequality. This example also illustrates the fact that the lattice distance is the minimum number of operations needed to change one flat into another. Here, moving from f to g requires two operations: firstly inserting an element from outside f to obtain $\mathrm{GF}(q)^n$, which has a basis given by n elements of g and one from outside g; then deleting this element to obtain g. In this sense, inequality [6.6] illustrates the sensitivity of network coding to errors.

This "shortcut" method for passing from f to g is because we can change the basis of a flat (here, $\mathrm{GF}(q)^n$) without modifying it. This is illustrated in Figure 6.3 for $\mathcal{A}(3, 2)$. The errors (here, the insertion of the element from outside of f) are propagated by combination and give way to a message without any relation to the transmitted flat f. The propagation of errors in Figure 6.3 is revealing because it is this same element that is inserted and then deleted and this element belongs to neither f nor g.

$$f : \begin{pmatrix} 0 & 0 \\ 1 & 0 \end{pmatrix} \xrightarrow{\text{insertion}} E : \begin{pmatrix} 0 & 0 \\ 1 & 0 \\ 0 & 2 \end{pmatrix} \xrightarrow{\text{combination}} E : \begin{pmatrix} 0 & 1 \\ 1 & 1 \\ 0 & 2 \end{pmatrix} \xrightarrow{\text{deletion}} g : \begin{pmatrix} 0 & 1 \\ 1 & 1 \end{pmatrix}$$

Figure 6.3. *Propagation of an error via a combination with RANC*

6.4.2. *Matroid codes*

For every simple matroid $\mathcal{M}$, we will define a matroid code as a non-empty set of flats of the same rank, or equivalently as a subset of $\mathcal{F}_k$. The minimum (lattice) distance of a matroid code is given by the minimum distance between two distinct codewords; by [6.5], it is always even. A matroid code with minimum distance of $2d$ can correct ϵ deletions and δ insertions, provided that $\epsilon + \delta < d$. A matroid code for SAF is a constant weight code [BRO 90]; the matroid codes for RLNC are introduced in [KOE 08] with the name "constant dimension codes"; the matroid codes for RANC are codes on affine subspaces.

We will now derive bounds on matroid codes. The maximum cardinality of a matroid code on the flats with rank k of $\mathcal{M}$ with minimum distance $2d$ is denoted by $A_{\mathcal{M}}(k,d)$. If the rank of $\mathcal{M}$ is r, we have $A_{\mathcal{M}}(k,1) = N_k$ for every $0 \leq k \leq r$ and $d_{\mathcal{M}}(f,g) \leq 2\min\{k, r-k\}$ for every $f, g \in \mathcal{F}_k$, motivating the convention that $A_{\mathcal{M}}(k,d) = 1$ for every $d > \min\{k, r-k\}$.

The Johnson bounds have been described first for constant weight codes [JOH 62] and subsequently adapted for constant dimension codes in [XIA 09, BOS 09]. Proposition 6.6 generalizes these bounds for matroid codes by restriction to a submatroid with an inferior rank in two ways. Firstly, for every $e \in E$, the contraction of e from $\mathcal{M}$, written as $\mathcal{M}/e$, is the simple matroid with flats $\mathcal{F}(\mathcal{M}/e) = \{f \in \mathcal{F} : e \in f\}$ [OXL 06, Chapter 3]. The matroid $\mathcal{M}/e$ has rank $r-1$, and for every $f, g \in \mathcal{F}(\mathcal{M}/e)$, $\mathrm{rk}_{\mathcal{M}/e}(f) = \mathrm{rk}_{\mathcal{M}}(f) - 1$ and as such $d_{\mathcal{M}/e}(f,g) = d_{\mathcal{M}}(f,g)$. Secondly, for every hyperplane $h \in \mathcal{F}_{r-1}$, the restriction of $\mathcal{M}$ to h, written as $\mathcal{M}|h$, is the simple matroid with flats $\{f \in \mathcal{F} : f \subseteq h\}$ [OXL 06, section 1.3]. For all flats $f, g \subseteq h$, $\mathrm{rk}_{\mathcal{M}|h}(f) = \mathrm{rk}_{\mathcal{M}}(f)$ and as such $d_{\mathcal{M}|h}(f, g) = d_{\mathcal{M}}(f, g)$.

PROPOSITION 6.6. (Johnson bounds).– For every $\mathcal{M}$ and $0 \leq k \leq r$, H_k is used to denote the minimum number of hyperplanes containing a given flat with rank k. Then there are $e \in E$ and $h \in \mathcal{F}_{r-1}$ such that

$$A_{\mathcal{M}}(k, d) \leq \frac{N_1}{C_k} A_{\mathcal{M}/e}(k - 1, d), \qquad [6.7]$$

$$A_{\mathcal{M}}(k, d) \leq \frac{N_{r-1}}{H_k} A_{\mathcal{M}|h}(k, d). \qquad [6.8]$$

Proposition 6.7 is a generalization of the Singleton bound for constant dimension codes in [KOE 08].

PROPOSITION 6.7. (Singleton bound).– For every $\mathcal{M}$, $0 \leq k \leq r$, and every element $e \in E$, we have

$$A_{\mathcal{M}}(k, d) \leq A_{\mathcal{M}/e}(k, d - 1)$$

and as such

$$A_{\mathcal{M}}(k, d) \leq \min_{g \in \mathcal{F}_{d-1}} |\{f \in \mathcal{F}_{k+d-1} : g \subseteq f\}|.$$

The Gilbert and Hamming bounds can also be adapted for matroid codes (see [KOE 08] for RLNC), but they are of little interest.

6.4.3. *Matroid codes for SAF*

For SAF, a matroid code is a code in the Johnson scheme [DEL 98], an association scheme formed by the set of subsets that are of the same size of a given set. Every subset $f \in \mathcal{F}_k$ of $[q^n]$ can be represented by its characteristic vector $\mathbf{f} \in \mathrm{GF}(2)^{q^n}$, where $f_i = 1$ if and only if $i \in f$. Equally, we have

$$f = \mathrm{support}(\mathbf{f}).$$

The lattice distance between the two subsets f, g is simply the Hamming distance between their characteristic vectors:

$$d_S(f, g) = d_H(\mathbf{f}, \mathbf{g}).$$

As such, a matroid code for SAF can be seen as a code over binary vectors with length q^n and weight k, commonly known as a constant weight code. These codes have been widely studied because they are of significant theoretical interest and have a number of applications. See [BRO 90] for further details. Note that the lattice distance in S reaches equality in [6.6].

The traditional techniques for protecting against losses of packets include ARQ and its derivatives, fountain codes, and erasure correcting codes [NON 98, XU 02] based on Reed–Solomon codes. The constant weight code approach is interesting because the operations for error control are carried out only at the destinations. In contrast to ARQ-type schemes, it does not require any feedback and contrary to fountain codes, messages are only sent once without the need to combine packets. We will now show that the erasure correcting codes can be seen as a particular case of constant weight codes.

The lifting $I_S(\mathbf{M})$ of a matrix $\mathbf{M} \in \mathrm{GF}(q)^{q^l \times (n-l)}$ is (the subset of $[q^n]$ with a cardinality of q^l) obtained by adding the header i, encoded in the form of l symbols of $\mathrm{GF}(q)$, before the row m_i. Every matrix of $\mathrm{GF}(q)^{q^l \times (n-l)}$ can be seen as a vector of $\mathrm{GF}(q^{n-l})^{q^l}$, where each row corresponds to a vector coordinate, we call the number of distinct rows in two matrices their Hamming distance. We can easily show that the lifting preserves the Hamming distance: for all $\mathbf{M}, \mathbf{N} \in \mathrm{GF}(q)^{q^l \times (n-l)}$,

$$d_S(I_S(\mathbf{M}), I_S(\mathbf{N})) = 2d_H(\mathbf{M}, \mathbf{N}).$$

As such, if the matrices $\mathbf{M}$ form a Reed–Solomon code with length q^l over $\mathrm{GF}(q^{n-l})$ with minimum distance d, the lifting of this code has lattice distance $2d$. The erasure correcting codes follow this construction exactly. Liftings are not optimal codes when the number of packets k is relatively close to q^n (i.e. an exponential number of packets). In contrast, for practical values, liftings are good matroid codes for SAF. Finally, note that the decoding algorithms developed for these codes only consider packet losses while we have seen that these codes in theory can also correct injections and packet errors.

6.5. Matroid codes for network coding

6.5.1. *Rank metric codes*

Rank metric codes are codes on matrices where the distance between two matrices is equal to the rank of their difference. They are explored in [DEL 78] for their theoretical interest because the set of matrices with the rank metric forms an association scheme, which is highly similar to the Hamming scheme. As such, as we will see, rank metric codes have characteristics similar to that of Hamming metric codes. They are also independently examined in [GAB 85] and [ROT 91] for correcting bidimensional errors in data storage devices. If the data are stored in the form of a two-dimensional array, then the corruption of a row or column can be seen as a rank one error. Rank metric codes have been proposed for a number of applications such as public-key cryptography [GAB 91], space-time coding [LUS 03], or even coding for channels with intersymbol interference [DUS 08].

More precisely, a rank metric code is a subset of matrices in $GF(q)^{k \times \nu}$. The distance between two matrices is the rank distance, defined as

$$d_{\mathrm{R}}(\mathbf{M}, \mathbf{N}) = \mathrm{rank}(\mathbf{M} - \mathbf{N}),$$

where rank denotes the number of linearly independent rows (or, equally, columns). Since matrix transposition preserves the rank, it also preserves the rank distance. As such, we can restrict ourselves to the situation where $k \geq \nu$, that is where the matrices have more rows than columns.

The Singleton bound indicates that a rank metric code in $\mathrm{GF}(q)^{k \times \nu}$ with minimum distance d contains at most $q^{k(\nu-d+1)}$ codewords. A class of codes attaining this bound, explored separately in [DEL 78, GAB 85, ROT 91] and known as Gabidulin codes, is given by analogs of Reed–Solomon codes. We review their construction in the following.

Firstly, the vector space $\mathrm{GF}(q)^k$ can be seen as the extension field $\mathrm{GF}(q^k)$ and therefore a Gabidulin code is normally defined as a code with length ν over $\mathrm{GF}(q^k)$. A linearized polynomial is a polynomial with coefficients in $\mathrm{GF}(q^k)$ with the form

$$p(x) = \sum_{i=0}^{k-1} p_i x^{q^i}.$$

The q-degree of $p(x)$ (the logarithm in base q of its degree) is written as $\deg(p)$; the q-degree of the zero polynomial is 0 by convention. A linearized polynomial is linear: $p(x + \lambda y) = p(x) + \lambda p(y)$ for every $x, y \in \mathrm{GF}(q^k)$ and every $\lambda \in \mathrm{GF}(q)$; conversely, every linear transformation of $\mathrm{GF}(q^k)$ can be expressed as a linearized polynomial. The roots of $p(x) \neq 0$ therefore form a subspace in $\mathrm{GF}(q^k)$ with dimension no more than $\deg(p)$. A Gabidulin code is therefore defined as follows: given $B = \{b_1, b_2, \ldots, b_\nu\}$ a set of linearly independent elements of $\mathrm{GF}(q^k)$, then $\mathcal{G}(B, d)$ is the set of evaluations on B of all the linearized polynomials with degree at most of $\nu - d$:

$$\mathcal{G}(B, d) = \{(p(b_1), p(b_2), \ldots, p(b_\nu)) \in \mathrm{GF}(q^k)^\nu : \deg(p) \leq \nu - d\}.$$

It is clear that this code has minimum rank distance d, and that its cardinality is exactly $q^{k(\nu-d+1)}$. More precisely, the generator matrix of $\mathcal{G}(B,d)$ is given by

$$\begin{pmatrix} b_1 & b_2 & \dots & b_\nu \\ b_1^q & b_2^q & \dots & b_\nu^q \\ \vdots & \vdots & \ddots & \vdots \\ b_1^{q^{\nu-d}} & b_2^{q^{\nu-d}} & \dots & b_\nu^{q^{\nu-d}} \end{pmatrix},$$

which illustrates the analogy between these codes and Reed–Solomon codes. By extension, the transposition of every Gabidulin code is also called a Gabidulin code.

Linearized polynomials equipped with composition form a non-commutative Euclidegan domain. Due to this property, a number of algorithms for Reed–Solomon codes based on polynomials can be adapted for linearized polynomials to obtain decoding algorithms for Gabidulin codes. Among these examples, we should mention the extended Euclidean [GAB 85], Peterson–Gorenstein–Zierler [ROT 91], Berlekamp–Massey [RIC 04], and Welch–Berlekamp [LOI 05] algorithms. However, no analog of the Guruswami–Sudan list-decoding algorithm [GUR 99] for Gabidulin codes is currently known.

6.5.2. *Matroid codes for RLNC*

A matroid code for RLNC is, by definition, a code on the linear subspaces of $GF(q)^n$ with dimension k or, in other words, a subset of all the subspaces of the same size, generally known as Grassmannian, which form an association scheme [DEL 76]. The codes on the Grassmannian, called constant dimension codes, are studied in further detail in [CHI 87, SCH 02] before their rediscovery by [KOE 08].

The lattice distance for $\mathcal{L}$ is called subspace distance, and reaches equality in [6.6]. In addition, for every linear subspace

$f \in F_k$, its dual $f^{\perp}$ is a subspace with a dimension of $n - k$. Considering the dual preserves the subspace distance, we have

$$d_{\mathcal{L}}(f^{\perp}, g^{\perp}) = d_{\mathcal{L}}(f, g)$$

for every $f, g \in F_k$. Therefore, $A_{\mathcal{L}}(k, d) = A_{\mathcal{L}}(n - k, d)$ and we subsequently suppose that $k \leq \frac{n}{2}$.

A highly practical and elegant construction for constant dimension codes is the linear lifting of rank metric codes. Remember that the linear lifting of a matrix $\mathbf{M} \in \mathrm{GF}(q)^{k \times (n-k)}$ is the row space of $(\mathbf{I}_k | \mathbf{M})$. The efficiency of linear liftings is based on two main ideas. Firstly, the major part (more than K_q in proportion) of subspaces with dimension k are liftings. Secondly, linear lifting preserves the rank distance. For all matrices $\mathbf{M}, \mathbf{N} \in \mathrm{GF}(q)^{k \times (n-k)}$,

$$d_{\mathcal{L}}(I_{\mathcal{L}}(\mathbf{M}), I_{\mathcal{L}}(\mathbf{N})) = 2d_{\mathrm{R}}(\mathbf{M}, \mathbf{N}).$$

As such, the complex problem of designing codes in the Grassmanian reduces to (at a minimum cost) using a rank metric code. In particular, the linear lifting of a Gabidulin code in $\mathrm{GF}(q)^{k \times (n-k)}$ and with minimum rank distance d is a matroid code for $\mathcal{L}$ with minimum distance $2d$ and cardinality $q^{(n-k)(k-d+1)}$. Decoding in the subspace distance can be carried out using the algorithm in [KOE 08]. The description of this algorithm is too long to be included here. Note that it is only based on the Sudan bivariational interpolation method [SUD 97]. Complexity is in the order of $O(n^2)$ operations over $\mathrm{GF}(q^k)$.

We will now demonstrate that linear liftings of Gabidulin codes are quasi-optimal. The Singleton bound offers the most elementary argument: every linear subspace with dimension $d - 1$ is contained in exactly $\begin{bmatrix} n-d+1 \\ k-d+1 \end{bmatrix}$ linear subspaces with dimension $n - k + d - 1$. We therefore obtain

$$A_{\mathcal{L}}(k, d) = A_{\mathcal{L}}(n-k, d) \leq \begin{bmatrix} n - d + 1 \\ k - d + 1 \end{bmatrix} < K_q^{-1} q^{(n-k)(k-d+1)}. \quad [6.9]$$

The Johnson bounds improve the Singleton bound somewhat [XIA 09].

Since [KOE 08], linear subspace codes have been intensely studied over three main areas of research:

– Firstly, [SIL 08] introduces a transmission model based on the rank metric and proposes a decoding algorithm based on the generalized Berlekamp–Massey algorithm. It should correct not only errors but also deletions and "deviations". This algorithm has been adapted for hardware implementation and is studied in [CHE 12].

– Secondly, it is easy to demonstrate that liftings are not optimal constant dimension codes, which open the way for code constructions that are larger than the liftings of Gabidulin codes. Until now, efforts have only taken the shape of timid advances by [ETZ 09, SIL 10a, SKA 10]. In any case, this area seems to be interesting only from a theoretical viewpoint: according to [6.9], the improvement in rate is only minimal.

– Thirdly, [SIL 09] proposes using variable dimension codes, that is codes on all the linear subspaces. This allows us to use not only codes with odd minimal subspace distances, but also a different metric called the injection distance, which is better adapted to scenarios where an adversary is present on the network. Unfortunately, the maximum size for variable dimension codes is very close to that of constant dimension codes for $k = \lfloor \frac{n}{2} \rfloor$ [GAD 10b].

6.5.3. *Matroid codes for RANC*

We will now examine codes on affine subspaces. By definition, we have $A_A(k, 1) = N_k = q^{n-k+1} \begin{bmatrix} n \\ k-1 \end{bmatrix} \geq q^{k(n-k+1)}$

for every $0 \leq k \leq n+1$. Since the affine geometry $\mathcal{A}(q,n)$ is a submatroid of the projective geometry $\mathcal{L}(q, n+1)$, we have

$$A_{\mathcal{A}(q,n)}(k,d) \leq A_{\mathcal{L}(q,n+1)}(k,d) < K_q^{-1} q^{\min\{(n-k+1)(k-d+1),k(n-k-d+2)\}}.$$

[6.10]

As we will see, this bound is tight up to the K_q^{-1} term. Nevertheless, we can obtain a stricter bound by applying the Johnson bound in Proposition 6.8 to codes on affine subspaces. We have $\mathcal{A}(q,n)/\mathbf{v} \cong \mathcal{L}(q,n)$ for every point $\mathbf{v} \in \mathrm{GF}(q)^n$ and $\mathcal{A}(q,n)|h \cong \mathcal{A}(q, n-1)$ for every affine hyperplane $h \subseteq \mathrm{GF}(q)^n$.

PROPOSITION 6.8. Bounds on matroid codes for RANC.– For every $2 \leq k \leq n-1$ and $2 \leq d \leq \min\{k, n-k+1\}$,

$$A_{\mathcal{A}(q,n)}(k,d) \leq q^{n-k+1} A_{\mathcal{L}(q,n)}(k-1,d), \qquad [6.11]$$

$$A_{\mathcal{A}(q,n)}(k,d) \leq \left\lfloor q\frac{q^n-1}{q^{n-k+1}-1} A_{\mathcal{A}(q,n-1)}(k,d) \right\rfloor \qquad [6.12]$$

$$\leq \left\lfloor q\frac{q^n-1}{q^{n-k+1}-1} \left\lfloor q\frac{q^{n-1}-1}{q^{n-k}-1} \cdots \left\lfloor q\frac{q^{k+d-1}-1}{q^d-1} \right\rfloor \cdots \right\rfloor \right\rfloor.$$

Note that the first Johnson bound in [6.11] is best for $2k \leq n+1$, while the second Johnson bound in [6.12] is best for $2k \geq n+1$. In addition, the Singleton bound for codes on affine subspaces is weaker than [6.11].

We will now design a class of quasi-optimum codes for RANC based on affine liftings of Gabidulin codes. Remember that for every matrix $\mathbf{M} \in \mathrm{GF}(q)^{k \times (n-k+1)}$, the affine lifting of $\mathbf{M}$, denoted by $I_A(\mathbf{M}) \in \mathcal{F}_k$, is the closure of the rows of $(\mathbf{I}'_k|\mathbf{M})$, where $\mathbf{I}'_k = (\mathbf{0}|\mathbf{I}_{k-1})^T \in \mathrm{GF}(q)^{k \times (k-1)}$. Note that

$$\mathrm{rk}(I_A(\mathbf{M})) = \mathrm{rank}(\mathbf{1}|\mathbf{I}'_k|\mathbf{M}) = k$$

and I_A sends $GF(q)^{k \times (n-k+1)}$ to $\mathcal{F}_k$. The affine lifting preserves the rank distance between the matrices. For every $\mathbf{M}, \mathbf{N} \in GF(q)^{k \times (n-k+1)}$,

$$d_A(I_A(\mathbf{M}), I_A(\mathbf{N})) = 2d_R(\mathbf{M}, \mathbf{N}).$$

Let C be a Gabidulin code on $GF(q)^{k \times (n-k+1)}$ with minimum rank distance d. Therefore, its affine lifting $I_A(C) = \{I_A(\mathbf{C}) : \mathbf{C} \in C\}$ is a matroid code of $\mathcal{A}(q, n)$ with minimum distance d and cardinality $q^{\min\{(n-k+1)(k-d+1), k(n-k-d+2)\}}$. In combination with [6.10], we obtain

$$A_{\mathcal{A}(q,n)}(k, d) \sim A_{\mathcal{L}(q,n+1)}(k, d).$$

As such, RANC uses codes with a similar cardinality to the codes used for RLNC for packets that are longer by one symbol. This improvement is clearly illustrated by the definition of affine lifting, which removes the first column of the identity matrix used for linear lifting. The gain in performance in [6.2] identified for the errorless case is therefore preserved when the packets are encoded to protect against errors.

Furthermore, we will demonstrate that this gain does not involve any increase in complexity for decoding. For every affine subspace $f \in \mathcal{F}(\mathcal{A}(q, n))$ seen as the closure of the rows of the matrix $\mathbf{N} \in GF(q)^{k \times n}$, let $\bar{f} \in \mathcal{F}(\mathcal{L}(q, n+1))$ be the linear subspace of $GF(q)^{n+1}$ with dimension k generated by the rows of

$$\left(1 - \sum_{i=0}^{k-2} \mathbf{col}_i \middle| \mathbf{N} \right) \in GF(q)^{k \times (n+1)},$$

where $\mathbf{col}_i$ is the ith column of $\mathbf{N}$.

PROPOSITION 6.9.– For every affine subspace $f \in \mathcal{F}(\mathcal{A}(q, n))$ and every matrix $\mathbf{C} \in GF(q)^{k \times (n-k)}$,

$$d_A(f, I_A(\mathbf{C})) = d_{\mathcal{L}}(\bar{f}, I_{\mathcal{L}}(\mathbf{C})).$$

According to proposition 6.9, decoding f for the affine lifting of a code is equivalent to decoding $\bar{f}$ for the linear lifting of the same code. As such, the decoding algorithm for the affine lifting of a Gabidulin code follows two stages. Firstly, add the column $1 - \sum_{i=0}^{k-2} \text{col}_i$ to the received matrix N, and then apply the algorithm from [KOE 08] for the row space of the extended matrix. It is clear that complexity does not increase when we consider the affine lifting in relation to the linear lifting of the same code.

To summarize the concepts examined throughout this chapter, the implementation scheme for RANC is illustrated in Figure 6.4 for $2k \leq n + 1$.

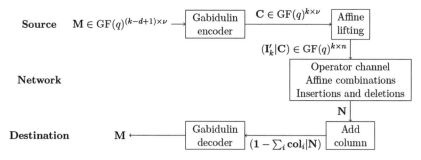

Figure 6.4. *Implementation scheme for RANC, where* $\nu = n - k + 1$

6.6. Conclusion

The concepts examined in this chapter open up a number of possibilities. Firstly, the proposed model is based on deliberately simple hypotheses that do not perhaps reflect the reality of the network considered. We therefore need to study how certain features of the network can be incorporated into the model. Secondly, different matroids can offer different compromises between the number of possible combinations and the data output. In addition, linear network coding is not optimal in situations with multiple sources [RII 04] and therefore non-representable matroids could offer a higher throughput than RANC or RLNC in this case. Thirdly, there are a number of ways of combining matroids, which can offer the opportunity to use several protocols on the network together or offer unequal levels of protection against errors.

6.7. Bibliography

[BOS 09] BOSSERT M., GABIDULIN E.M., "One family of algebraic codes for network coding", *Proceedings of IEEE International Symposium on Information Theory*, Seoul, South Korea, pp. 2863–2866, June–July 2009.

[BRO 90] BROUWER A.E., SHEARER J.B., SLOANE N.J.A., SMITH W.D., "A new table of constant weight codes", *IEEE Transactions on Information Theory*, vol. 36, no. 6, pp. 1334–1380, November 1990.

[CAI 06] CAI N., YEUNG R.W., "Network error correction, part II: lower bounds", *Communications in Information and Systems*, vol. 6, no. 1, pp. 37–54, 2006.

[CHE 12] CHEN N., YAN Z., GADOULEAU M., WANG Y., SUTER B.W., "Rank metric decoder architectures for random linear network coding with error control", *IEEE Transactions on VLSI Systems*, vol. 20, no. 2, pp. 296–309, February 2012.

[CHI 87] CHIHARA L., "On the zeros of the Askey–Wilson polynomials with applications to coding theory", *SIAM Journal on Mathematical Analysis*, vol. 8, no. 1, pp. 191–207, January 1987.

[DEL 76] DELSARTE P., "Association schemes and t-designs in regular semilattices", *Journal of Combinatorial Theory A*, vol. 20, no. 2, pp. 230–243, March 1976.

[DEL 78] DELSARTE P., "Bilinear forms over a finite field, with applications to coding theory", *Journal of Combinatorial Theory A*, vol. 25, no. 3, pp. 226–241, November 1978.

[DEL 98] DELSARTE P., LEVENSHTEIN V.I., "Association schemes and coding theory", *IEEE Transactions on Information Theory*, vol. 44, no. 6, pp. 2477–2504, October 1998.

[DEZ 92] DEZA M., "Matroid applications", *Perfect Matroid Designs*, Cambridge University Press, pp. 54–72, 1992.

[DUS 08] DUSAD S., DIGGAVI S.N., CALDERBANK A.R., "Embedded rank distance codes for ISI channels", *IEEE Transactions on Information Theory*, vol. 54, no. 11, pp. 4866–4886, November 2008.

[ETZ 09] ETZION T., SILBERSTEIN N., "Error-correcting codes in projective spaces via rank-metric codes and ferrers diagrams", *IEEE Transactions on Information Theory*, vol. 55, no. 7, pp. 2909–2919, July 2009.

[GAB 85] GABIDULIN E.M., "Theory of codes with maximum rank distance", *Problems of Information Transmission*, vol. 21, no. 1, pp. 1–12, January 1985.

[GAB 91] GABIDULIN E.M., PARAMONOV A.V., TRETJAKOV O.V., "Ideals over a non-commutative ring and their application in cryptology", *Proceedings of Eurocrypt*, Brighton, UK, pp. 482–489, April 1991.

[GAD 10a] GADOULEAU M., GOUPIL A., "Binary codes for packet error and packet loss correction in store and forward", *Proceedings of International ITG Conference on Source and Channel Coding*, Siegen, Germany, January 2010.

[GAD 10b] GADOULEAU M., YAN Z., "Packing and covering properties of subspace codes for error control in random linear network coding", *IEEE Transactions on Information Theory*, vol. 56, no. 5, pp. 2097–2108, May 2010.

[GAD 11] GADOULEAU M., GOUPIL A., "A matroid framework for non-coherent random network communications", *IEEE Transactions on Information Theory*, vol. 57, no. 2, pp. 1031–1045, February 2011.

[GUR 99] GURUSWAMI V., SUDAN M., "Improved decoding of Reed–Solomon and algebraic-geometry codes", *IEEE Transactions on Information Theory*, vol. 45, no. 6, pp. 1757–1767, September 1999.

[HO 06] HO T., MÉDARD M., KÖTTER R., KARGER D.R., EFFROS M., SHI J., LEONG B., "A random linear network coding approach to multicast", *IEEE Transactions on Information Theory*, vol. 52, no. 10, pp. 4413–4430, October 2006.

[JOH 62] JOHNSON S.M., "A new upper bound for error-correcting codes", *IRE Transactions on Information Theory*, vol. 8, no. 3, pp. 203–207, April 1962.

[KOE 08] KOETTER R., KSCHISCHANG F.R., "Coding for errors and erasures in random network coding", *IEEE Transactions on Information Theory*, vol. 54, no. 8, pp. 3579–3591, August 2008.

[LOI 05] LOIDREAU P., "A Welch–Berlekamp like algorithm for decoding gabidulin codes", *Proceedings of International Workshop on Coding and Cryptography*, Bergen, Norway, pp. 36–45, March 2005.

[LUS 03] LUSINA P., GABIDULIN E.M., BOSSERT M., "Maximum rank distance codes as space-time codes", *IEEE Transactions on Information Theory*, vol. 49, no. 10, pp. 2757–2760, October 2003.

[NON 98] NONNENMACHER J., BIERSACK E.W., TOWSLEY D., "Parity-based loss recovery for reliable multicast transmission", *IEEE/ACM Transactions on Networking*, vol. 6, no. 4, pp. 349–361, August 1998.

[OXL 06] OXLEY J.G., *Matroid Theory*, Oxford University Press, 2006.

[RIC 04] RICHTER G., PLASS S., "Error and erasure decoding of rank-codes with a modified Berlekamp–Massey algorithm", *Proceedings of ITG Conference on Source and Channel Coding*, Erlangen, Germany, pp. 203–211, January 2004.

[RII 04] RIIS S., "Linear versus non-linear Boolean functions in network flow", *Proceedings CISS*, Princeton, USA, March 2004.

[ROT 91] ROTH R.M., "Maximum-rank array codes and their application to crisscross error correction", *IEEE Transactions on Information Theory*, vol. 37, no. 2, pp. 328–336, March 1991.

[SCH 02] SCHWARTZ M., ETZION T., "Codes and anticodes in the Grassmann graph", *Journal of Combinatorial Theory A*, vol. 97, no. 1, pp. 27–42, January 2002.

[SIL 10a] SILBERSTEIN N., ETZION T., "Large constant-dimension codes and lexicodes", *Proceedings of Algebraic Combinatorics and Applications*, Thurnau, Germany, April 2010.

[SIL 08] SILVA D., KSCHISCHANG F.R., KÖTTER R., "A rank-metric approach to error control in random network coding", *IEEE Transactions on Information Theory*, vol. 54, no. 9, pp. 3951–3967, September 2008.

[SIL 09] SILVA D., KSCHISCHANG F.R., "On metrics for error correction in network coding", *IEEE Transactions on Information Theory*, vol. 55, no. 12, pp. 5479–5490, December 2009.

[SIL 10b] SILVA D., KSCHISCHANG F.R., KÖTTER R., "Communication over finite-field matrix channels", *IEEE Transactions on Information Theory*, vol. 56, no. 3, pp. 1296–1305, March 2010.

[SKA 10] SKACHEK V., "Recursive code construction for random networks", *IEEE Transactions on Information Theory*, vol. 56, no. 3, pp. 1378–1382, March 2010.

[SUD 97] SUDAN M., "Decoding of ReedSolomon codes beyond the error-correction diameter", *Proceedings of Allerton Conference on Communication, Control and Computing*, Monticello, USA, pp. 215–224, September–October 1997.

[XIA 09] XIA S.-T., FU F.-W., "Johnson type bounds on constant dimension codes", *Designs, Codes and Cryptography*, vol. 50, no. 2, pp. 163–172, February 2009.

[XU 02] XU Y., ZHANG T., "Variable shortened-and-punctured Reed–Solomon codes for packet loss protection", *IEEE Transactions on Broadcasting*, vol. 48, no. 3, pp. 237–245, September 2002.

[YEU 06] YEUNG R.W., CAI N., "Network error correction, part I: basic concepts and upper bounds", *Communications in Information and Systems*, vol. 6, no. 1, pp. 19–36, 2006.

Chapter 7

Joint Network-Channel Coding for the Semi-Orthogonal MARC: Theoretical Bounds and Practical Design

7.1. Introduction

Network coding, initially proposed by Ahlswede, Cai, Li, and Yeung in [AHL 00], is a novel and powerful approach where intermediate nodes in a network are allowed to not only route but also perform algebraic operations on the incoming data flows. By applying network coding in multicast transmission with a single source and lossless links, the authors of [AHL 00] have proved that the network throughput could achieve the min-cut max-flow capacity between the source and the sinks. This remarkable result has motivated further theoretical and practical research to extend network coding to wireless media, which, albeit lossy, greatly facilitate

Chapter written by Atoosa HATEFI, Antoine O. BERTHET and Raphaël VISOZ.

its application, as broadcast is guaranteed at no cost. Among the first contributions are [KAT 05, WU 05, KAT 06], which focus again on lossless channels, and [GUO 07], which considers lossy erasure channels in sensor networks.

The multiple-access relay channel (MARC) [KRA 00, SAN 04] is the simplest multiterminal network for physical layer network coding. It consists of M sources that want to communicate to a destination in the presence of a relay. In this chapter, we retain a restrictive version, namely, MARC with time-division-based half-duplex relaying. There are essentially two ways of performing network coding: separate network channel coding (SNCC) and joint network channel coding (JNCC). In SNCC, channel coding is performed locally and separately for each transmission to transform the noisy channels into erasure-based links. On the network layer, network coding is performed for the erasure-based networks, which are provided by the lower layers [CHE 06, BAO 08]. SNCC requires separate network channel decoding (SNCD) at the destination, in which channel decoding is first performed at the physical layer and outputs the estimates to the network decoder. As the network layer typically uses the addition modulo 2 (XOR) operation to perform network coding, it can retrieve the M messages, if at least M out of $M + 1$ channel output estimates are error-free. However, in JNCC, we exploit the redundancy of the network code to support the channel code, which can finally improve the coding gain of the system. A joint network channel decoding (JNCD) is then performed at the destination, in which soft information between the network decoder and the channel decoders is exchanged.

7.1.1. *Related work*

To the best of the authors' knowledge, Hausl *et al.* were among the first to describe efficient JNCC based on low-density parity check (LDPC) codes [HAU 05] or turbo codes

[HAU 06]. Common to this set of contributions are the hypotheses of (1) joint decode and forward strategy, that is the relay cooperates if and only if all the decoded messages are error-free; and (2) orthogonality between all the radio links. Concerning the second hypothesis, we have already pointed out that wireless media naturally offer broadcast without an additional cost. This intrinsic property comes at the price of signal superposition (i.e. interference) at all intermediate nodes and at the destination. To fight back this impediment, orthogonal medium access (in time, frequency, or code space) is often assumed in cooperative communications. If orthogonality greatly simplifies the design of JNCC/JNCD and the performance analysis, it also substantially reduces the spectral efficiency of the proposed systems. Indeed, from an information-theoretic point of view, orthogonal multiple-access is, in general, not optimal for the slow fading (quasi-static) channel, although it may be close to optimal at a very low signal-to-noise ratio (SNR). Furthermore, in [HAU 05], the authors assume error-free source-to-relay links. To justify this hypothesis, we could imagine a restrictive communication scenario where the relay is very close to, and in line-of-sight with, the two sources. But even in this case, some decoding errors would occur at the relay, since, in practice, constituent codes used on point-to-point links are never perfect. Furthermore, the above code designs do not guarantee full diversity. More recently, JNCC based on LDPC codes were presented in [DUY 10] in the case of orthogonal links and error-free source-to-relay links, where full diversity is guaranteed by construction. However, it is not generic in terms of coding choice and the number of sources. Moreover, its coding gain decreases enormously for the case of error-prone source-to-relay links, even if its full diversity structure is maintained. Another approach guaranteeing full diversity is proposed in [WAN 08] assuming error-prone links. The authors in [WAN 08] employ link-adaptive regenerative (LAR) relaying scheme in which the detected symbols at the relay are

synchronized. All links of the network are subject to slow fading and are prone to errors. Neither the sources nor the relay when it transmits has channel state information, for example by means of feedback channels. The relay, when it listens and decodes, and the destination have perfect channel state information.

As a first contribution, we propose a new class of MARC that we call semi-orthogonal MARC (SOMARC) and is defined as follows: (1) independent sources communicate with a single destination in the presence of a relay; (2) the relay is half-duplex and applies a selective decode and forward (SDF) relaying strategy, that is it forwards only a deterministic function of the messages that it can decode without errors; and (3) the sources are allowed to transmit simultaneously during the listening phase of the relay, but are constrained to remain silent during its transmission phase. Allowing collisions at the relay and the destination renders the reality of wireless environments and leverages better the broadcast nature of the radio channel than the orthogonal MARC (OMARC). Furthermore, the proposed SDF in SOMARC not only prevents the error propagation from the relay to the destination, but also decreases the individual Block Error Rate (BLER), that is the BLER for each source. This SDF approach is theoretically analyzed in [WOL 07] for the OMARC/SNCC, and is shown to have a good information outage probability, when the quality of source-to-relay links is poor. For the case of JNCC, the information outage probability of OMARC is analyzed in [DUY 10] conditional on the error-free source-to-relay links, and can be easily generalized to the case of selective relaying. While information-theoretic analysis of OMARC with selective relaying has provided insight into the behavior of the system, many issues need to be addressed, including the impact of the non-orthogonality and the multiple-access interference.

Based on a careful outage analysis, the SOMARC individual information outage probability (e.g. for S_1) is derived for both JNCC or SNCC. The individual information outage probability and the individual ϵ-outage capacity (e.g. for S1) are then numerically evaluated assuming independent Gaussian inputs or discrete independent identically uniformly inputs and compared with the ones of an OMARC at fixed energy budget per source (per available dimensions). As a second contribution, we propose practical JNCC designs for SOMARC that are flexible in terms of number of sources and Modulation Coding Scheme (MCS). Our designs are built on convolutional codes and turbo codes, and rely on advanced (iterative) joint detection and decoding receiver architectures. We further demonstrate that they also guarantee the full diversity in the sense that they achieve the same diversity gain as the single-user case. The rationale behind our code construction is as follows. In the large SNR regime and for the special case of one receive antenna, the outage probability of an M-user slow fading Multiple Access Channel (MAC) behaves as that of an orthogonal MAC. Besides, a necessary condition for JNCC to achieve full diversity is that it achieves full diversity over the Block Erasure Channel (BEC) defined as an abstraction of the original channel in which the fading gains belong to the set $\{0, \infty\}$. This corresponds to the large SNR regime (see [DUY 10] and the references therein). As a result, in the large SNR regime, the MACs at the relay and at the destination and the point-to-point channel from relay to destination turn into five independent BECs. We claim that our proposed JNCC schemes are full diversity since the BLER of each source decays as ϵ^2, where ϵ is the probability of each link to be in erasure.

7.1.3. *Chapter outline*

The remainder of the chapter is outlined as follows. In section 7.2, we introduce the system model. Section 7.3

is devoted to an outage analysis of SOMARC/JNCC and SOMARC/SNCC. In section 7.4, we describe each aspect of our proposed JNCC/JNCD, including the relaying function. In section 7.5, we briefly discuss the SNCC/SNCD (used as a reference). Different scenarios are presented and compared in section 7.6. Some conclusions are drawn in section 7.7.

7.1.4. *Notation*

In the following, we use boldface letters to denote vector and matrices. Matrices are represented by capital letters. Let $\mathbf{A}$ be a matrix with ith row $\boldsymbol{a}^i$ and jth column $\boldsymbol{a}_j$, entry (i, j) is denoted $a_{i,j}$ or equivalently $[\mathbf{A}]_{i,j}$. The n-square identity matrix is denoted by $\mathbf{I}_n$. $\mathbf{0}_n$ and $\mathbf{1}_n$ stand for n-tuples of zeros and ones, respectively. diag $\{.\}$ denotes diagonal operator on square matrices. Superscript † indicates complex conjugate matrix transpose. $\mathbf{x} \sim p(\mathbf{x})$ means that the random vector $\mathbf{x}$ follows the probability distribution function $p(\mathbf{x})$. $\mathbf{x} \sim \mathcal{CN}(\boldsymbol{\mu}, \boldsymbol{\Sigma})$ means that $\mathbf{x}$ is a circularly symmetric complex Gaussian random vector with mean $\boldsymbol{\mu}$ and covariance $\boldsymbol{\Sigma}$. Following [GRA 89], we adopt the Iverson's convention of behavioral theory, that is if P is a predicate (Boolean proposition) involving some set of variables, then $[P] = 1$ if P is true and $[P] = 0$ otherwise.

7.2. System model

The M statistically independent sources $S_1, \dots, S_M$ want to communicate with the destination D in the presence of a relay R. To create virtual uplink multiple-input multiple-output (MIMO) channels and to benefit from spatial multiplexing gains, we assume that the relay R and the destination D are equipped with N_R and N_D receive antennas. We consider that the baud rate of the sources and relay is $D = 1/T_s$ and the overall transmission time is fixed to T, thus the number of available channel uses to be shared between the sources and the relay is $N = DT$. We consider the case of Nyquist rate and cardinal sine transmission pulse shape, that is $N = DT$ is the total number of available complex dimensions and D is the total bandwidth of the system. Our channel models are inspired by the following assumptions: (1) the delay spreads of the radio channels from the sources to the relay and the destination as well as from the relay to the destination are much lower than T_s ensuring no frequency selectivity; (2) the coherence time of all the aforementioned radio channels are assumed to be much larger than T. The semi-orthogonal transmission protocol is considered. The N available channel uses are divided into two successive time slots corresponding to the listening phase of the relay, say $N_1 = \alpha N$ channel uses, and to the transmission phase of the relay, say $N_2 = \bar{\alpha} N$ channel uses, with $\alpha \in [0,1]$ and $\bar{\alpha} = 1 - \alpha$. Each source i broadcasts its messages $\mathbf{u}_i \in \mathbb{F}_2^K$ of K information bits under the form of a modulated sequence during the first transmission phase. Without loss of generality, the modulated sequences are chosen from the complex codebooks ζ_i of rate $K/(\alpha N)$ and take the form of sequences $\mathbf{x}_i \in \zeta_i \subset \mathscr{X}_i^{\alpha N}$, $i \in \{1, \ldots, M\}$, where $\mathscr{X}_i \subset \mathbb{C}$ denote a complex signal set of cardinality $|\mathscr{X}_i| = 2^{q_i}$, with energy normalized to unity. The corresponding received signals at the relay and destination are expressed as

$$\mathbf{y}_{R,k}^{(1)} = \sum_{i=1}^{M} \sqrt{P_{iR}} \mathbf{h}_{iR} x_{i,k} + \mathbf{n}_{R,k}^{(1)} \qquad [7.1]$$

$$\mathbf{y}_{D,k}^{(1)} = \sum_{i=1}^{M} \sqrt{P_{iD}} \mathbf{h}_{iD} x_{i,k} + \mathbf{n}_{D,k}^{(1)} \qquad [7.2]$$

for $k = 1, \ldots, \alpha N$. In [7.1] and [7.2], the channel fading vectors $\mathbf{h}_{iR} \in \mathbb{C}^{N_R}$, and $\mathbf{h}_{iD} \in \mathbb{C}^{N_D}$, $i \in \{1, \ldots, M\}$, are mutually independent, constant over the transmission of $\mathbf{x}_1, \ldots, \mathbf{x}_M$, and change independently from one transmission of sources to the next. The channel fading vectors $\mathbf{h}_{iR}$, $i \in \{1, \ldots, M\}$, are identically distributed (i.d.) following the probability density function (pdf) $\mathcal{CN}(\mathbf{0}_{N_R}, \mathbf{I}_{N_R})$. The channel fading vectors $\mathbf{h}_{iD}$, $i \in \{1, \ldots, M\}$, are i.d. following the pdf $\mathcal{CN}(\mathbf{0}_{N_D}, \mathbf{I}_{N_D})$. The additive noise vectors $\mathbf{n}_{R,k}^{(1)}$ and $\mathbf{n}_{D,k}^{(1)}$ are independent and follow the pdf $\mathcal{CN}(\mathbf{0}_{N_R}, N_0 \mathbf{I}_{N_R})$ and $\mathcal{CN}(\mathbf{0}_{N_D}, N_0 \mathbf{I}_{N_D})$, respectively. $P_{ij} \propto (d_{ij}/d_0)^{-\kappa} P_i$, $i \in \{1, \ldots, M\}$, $j \in \{R, D\}$, is the average receive energy per dimension and per antenna (in Joules/symbol), where d_{ij} is the distance between the transmitter and the receiver, d_0 is a reference distance, κ is the path loss coefficient, with values typically in the range of $[2, 6]$, and P_i is the transmit power (or energy per symbol) of the source i. Note that the shadowing could be included within P_{ij}. To fairly compare the performance with respect to α, we fix the total energy per available dimensions $N P_{0,i}$ (recall that N is the number of available dimensions or channel uses) spent by source i, that is $P_i = P_{0,i}/\alpha$. During the second phase, both sources are silent. The relay uses an SDF approach, which depends on the number of correctly decoded messages. Let $J = \{j_1, j_2, \ldots, j_{|J|}\}$, $|J| \leq M$, denote the set of message indices with cardinality $|J|$ that have been successfully decoded. If $J = \emptyset$, the relay remains silent. Otherwise, according to the number of correctly decoded messages and the chosen network coding scheme, it transmits a modulated sequence of the form $\mathbf{x}_R \in \mathscr{X}_R^{\bar{\alpha} N}$, where $\mathscr{X}_R \subset \mathbb{C}$ is a complex constellation of order $|\mathscr{X}_R| = 2^{q_R}$ with energy normalized to unity. The modulated sequence $\mathbf{x}_R$ is chosen such that $(\mathbf{x}_{j_1}, \ldots, \mathbf{x}_{j_{|J|}}, \mathbf{x}_R)$ is a codeword on message $(\mathbf{u}_{j_1}, \ldots, \mathbf{u}_{j_{|J|}})$ belonging to a codebook $\zeta_{J,R}$ of rate $|J| K/N$. The received signal at the destination is expressed as

$$\mathbf{y}_{D,k}^{(2)} = \theta \sqrt{P_{RD}} \mathbf{h}_{RD} x_{R,k} + \mathbf{n}_{D,k}^{(2)} \qquad [7.3]$$

for $k = 1, \ldots, \bar{\alpha} N$. In [7.3], the channel fading vector $\mathbf{h}_{RD} \in \mathbb{C}^{N_D}$ follows the pdf $\mathcal{CN}(\mathbf{0}_{N_D}, \mathbf{I}_{N_D})$, is independent of $\mathbf{h}_{iD}$, $i \in \{1, \ldots, M\}$, constant over the transmission of $\mathbf{x}_R$, and changes independently from one transmission of the relay to another. The additive noise vector $\mathbf{n}_{D,k}^{(2)}$ is independent of $\mathbf{n}_{R,k}^{(1)}$ and $\mathbf{n}_{D,k}^{(1)}$, and follows the pdf $\mathcal{CN}(\mathbf{0}_{N_D}, N_0 \mathbf{I}_{N_D})$. $P_{RD} \propto (d_{RD}/d_0)^{-\kappa} P_R$, with P_R being the transmit power of the relay, is the average receive power per dimension and per antenna at the destination. Here again, we fix the total energy per available dimensions $N P_{0,R}$ spent by the relay, that is $P_R = P_{0,R}/\bar{\alpha}$. The parameter θ is a discrete Bernoulli distributed random variable: $\theta = 1$ if the relay successfully decodes at least one source message, and $\theta = 0$ otherwise.

Concerning the relay functionality, we distinguish the two cases of JNCC and SNCC:

– JNCC: the relay interleaves each message $\mathbf{u}_j$, $j \in J$, by π and applies a function $\Theta_{R,|J|}$

$$\Theta_{R,|J|} : \underbrace{\mathbb{F}_2^K \times \mathbb{F}_2^K \times \ldots \times \mathbb{F}_2^K}_{|J|} \to \mathbb{C}^{\bar{\alpha}N} \qquad [7.4]$$

to obtain the modulated sequence $\mathbf{x}_R$. In general, the function $\Theta_{R,|J|}$ is not a bijection on the interleaved correctly decoded messages. In practice, the relay would add some in-band signaling to make the destination aware of the set J. Finally, the relay signal, together with the source signals, forms a distributed joint network-channel codebook. The block diagram of the system model is depicted in Figure 7.1 for the case of $M = 2$.

– SNCC: the relay sums the correctly decoded messages using XOR (or the addition in $\mathbb{F}_2$). The resulting vector $\mathbf{u}_R \in \mathbb{F}_2^K$ is then mapped to $\mathbf{x}_R$ using the codebook ζ_R of rate $K/\bar{\alpha}N$.

In the rest of the chapter, for the sake of notational simplicity, we consider $M = 2$ sources that transmit with an overall spectral efficiency $r = K/N$. For the specific case of SNCC, we can associate the same spectral efficiency r to the relay transmission.

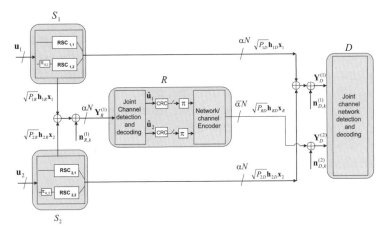

Figure 7.1. *System model (relay cooperates)*

The generalization to the cases of $M > 2$ sources is straightforward.

7.3. Information-theoretic analysis

The SOMARC breaks down into two MACs at the relay and destination, and one classical single user channel from the relay to the destination thanks to the SDF relaying function. Thus, its capacity region is a perfectly known conditional on a given channel state $\mathbf{H} = \begin{bmatrix} \mathbf{h}_{1R} & \mathbf{h}_{2R} & \mathbf{h}_{1D} & \mathbf{h}_{2D} & \mathbf{h}_{RD} \end{bmatrix}$. Let us define the independent input random variables $x_1 \sim p(x_1)$, $x_2 \sim p(x_2)$, and $x_R \sim p(x_R)$, and the associated independent output random vectors $\mathbf{y}_D^{(1)}$, $\mathbf{y}_D^{(2)}$, and $\mathbf{y}_R^{(1)}$ whose channel transition conditional pdfs are $p(\mathbf{y}_D^{(1)} \mid x_1, x_2, \mathbf{H}) = \mathcal{CN}(\sqrt{P_{1D}}\mathbf{h}_{1D}x_1 + \sqrt{P_{2D}}\mathbf{h}_{2D}x_2, N_0\mathbf{I}_{N_D})$, $p(\mathbf{y}_D^{(2)} \mid x_R, \mathbf{H}) = \mathcal{CN}(\sqrt{P_{RD}}\mathbf{h}_{RD}x_R, N_0\mathbf{I}_{N_D})$, and $p(\mathbf{y}_R^{(1)} \mid x_1, x_2, \mathbf{H}) = \mathcal{CN}(\sqrt{P_{1R}}\mathbf{h}_{1R}x_1 + \sqrt{P_{2R}}\mathbf{h}_{2R}x_2, N_0\mathbf{I}_{N_R})$. It follows that the mutual informations $I(x_1, x_2; \mathbf{y}_D^{(1)})$, $I(x_R; \mathbf{y}_D^{(2)})$, and $I(x_1, x_2; \mathbf{y}_R^{(1)})$ are perfectly defined by the pdfs $p(x_1)$, $p(x_2)$, $p(x_R)$, and the aforementioned channel transition probabilities. It is clear from our context that the mutual information conditional on any given channel state is maximized for the pdfs $p(x_1)$, $p(x_2)$, and $p(x_R)$ being circularly symmetric complex Gaussian pdfs. As a result, the latter pdfs minimize the information outage probabilities. However, in practice, $p(x_1)$, $p(x_2)$, and $p(x_R)$ are uniformly distributed pmfs (dirac comb pdfs) over the chosen constellation alphabets. That is why both cases are investigated in the following. We recall that in our analysis:

1) The theoretical bounds are derived conditional on both JNCC and SNCC.

2) The SDF relaying function is used under the hypothesis that all the links are prone to errors.

3) The sequences x_1, x_2, and x_R are the outcomes of independent discrete time i.i.d. processes whose associated pdfs are $p(x_1)$, $p(x_2)$, and $p(x_R)$ and their respective length is infinite ($N \rightarrow \infty$) such that the asymptotic equipartition property (AEP) holds.

4) The outage limit is either the individual information outage probability or the individual ϵ-outage capacity (e.g. for S_1).

These outage analyses are then used in section 7.6 to compare: (1) the performance of SOMARC/JNCC versus SOMARC/SNCC; (2) the performance of SOMARC/JNCC versus OMARC/JNCC; and (3) the BLER performance of practical designs versus the information outage probability, since the information outage probability turns out to be a tight estimate of the average BLER even for finite N [MAL 99, Chapter 10].

7.3.1. *Outage analysis of SOMARC/JNCC*

As the relay uses an SDF approach, an evaluation of the source-to-relay channel quality has to be processed first.

Let $\mathcal{E}_R(\mathbf{H})$ denote the outage event of the source-to-relay MAC conditional on $\mathbf{H}$. It corresponds to the case where the relay cannot decode both messages correctly, and can be expressed as

$$\mathcal{E}_R(\mathbf{H}) = \mathcal{E}_{R,1|2}(\mathbf{H}) \cup \mathcal{E}_{R,2|1}(\mathbf{H}) \cup \mathcal{E}_{R,1,2}(\mathbf{H}) \qquad [7.5]$$

where $\mathcal{E}_{R,i|j}(\mathbf{H})$, $i,j \in \{1,2\}$, $j \neq i$, is the outage event of S_i if the information of S_j is known, and $\mathcal{E}_{R,1,2}(\mathbf{H})$ is the outage event of both sources at the relay. The three possible outage events are then given by

$$\mathcal{E}_{R,i|j}(\mathbf{H}) = \left\{ \alpha I(x_i; \mathbf{y}_R^{(1)} \mid x_j) < r \right\} \qquad [7.6]$$

$$\mathcal{E}_{R,1,2}(\mathbf{H}) = \left\{ \alpha I(x_1, x_2; \mathbf{y}_R^{(1)}) < 2r \right\} \qquad [7.7]$$

When the outage event $\mathcal{E}_R(\mathbf{H})$ holds, in order to verify whether only one of the messages x_i can be successfully decoded or not, we define the following outage event

$$\mathcal{E}_{R,i}(\mathbf{H}) = \left\{ \alpha I(x_i; \mathbf{y}_R^{(1)}) < r \right\} \qquad [7.8]$$

in which the relay treats the signal x_j as interference. Thus, the relay outage events for the SDF approach can be summarized as follows: (1) in case of $\mathcal{Q}_R^{(1)}(\mathbf{H}) = \bar{\mathcal{E}}_R(\mathbf{H})$, which indicates the complement of the outage event $\mathcal{E}_R(\mathbf{H})$, the relay cooperates with both sources; (2) in case of $\mathcal{Q}_R^{(2)}(\mathbf{H}) = \mathcal{E}_R(\mathbf{H}) \cap \bar{\mathcal{E}}_{R,1}(\mathbf{H})$, the relay cooperates only with S_1; (3) in case of $\mathcal{Q}_R^{(3)}(\mathbf{H}) = \mathcal{E}_R(\mathbf{H}) \cap \bar{\mathcal{E}}_{R,2}(\mathbf{H})$, the relay cooperates only with S_2; (4) otherwise, in case of $\mathcal{Q}_R^{(4)}(\mathbf{H}) = \mathcal{E}_R(\mathbf{H}) \cap \mathcal{E}_{R,1}(\mathbf{H}) \cap \mathcal{E}_{R,2}(\mathbf{H})$, the relay does not cooperate. Now, depending on the relay transmission, we distinguish four outage events at the destination:

Case 1: The relay cooperates with both sources. The destination always receives the cooperative information from the relay during the second phase. Since the relay transmits over an orthogonal parallel channel with respect to the source-to-destination MAC, the outage at the destination occurs if the target rate exceeds the sum of the mutual informations of the two parallel channels. Let $\mathcal{E}_D^{(1)}(\mathbf{H})$ denote the outage event at the destination conditional on $\mathbf{H}$. It can be expressed as

$$\mathcal{E}_D^{(1)}(\mathbf{H}) = \mathcal{E}_{D,1|2}^{(1)}(\mathbf{H}) \cup \mathcal{E}_{D,2|1}^{(1)}(\mathbf{H}) \cup \mathcal{E}_{D,1,2}^{(1)}(\mathbf{H}) \qquad [7.9]$$

where

$$\mathcal{E}_{D,i|j}^{(1)}(\mathbf{H}) = \left\{ \alpha I(x_i; \mathbf{y}_D^{(1)} \mid x_j) + \bar{\alpha} I(x_R; \mathbf{y}_D^{(2)}) < r \right\} \qquad [7.10]$$

for $i, j \in \{1, 2\}$ and $j \neq i$, and

$$\mathcal{E}_{D,1,2}^{(1)}(\mathbf{H}) = \left\{ \alpha I(x_1, x_2; \mathbf{y}_D^{(1)}) + \bar{\alpha} I(x_R; \mathbf{y}_D^{(2)}) < 2r \right\} \qquad [7.11]$$

In [7.10], $\mathcal{E}_{D,i|j}^{(1)}(\mathbf{H})$, $i \in \{1, 2\}$, is the outage event of S_i if the information of S_j, $j \neq i$, is known. In this case, x_R can be considered a part of the codeword of S_i. Typically, this is the case when x_R is a codeword representing the XOR of the two source messages. The outage event in [7.11] corresponds to the constraint that the total throughput cannot exceed the sum of the mutual informations of (1) a point-to-point channel with the aggregate received signals of the two sources, and (2) the relay-to-destination channel. When $\mathcal{E}_D^{(1)}(\mathbf{H})$ holds, the destination cannot correctly decode both the source messages. As we are interested in calculating the outage event of S_1, we define the following event

$$\mathcal{E}_{D,1}^{(1)}(\mathbf{H}) = \left\{ \alpha I(x_1; \mathbf{y}_D^{(1)}) < r \right\} \qquad [7.12]$$

in which the destination treats the signal x_2 as interference. It is worth noting that the relay transmission in this case cannot help S_1 as it contains the interference from S_2. Thus, it is considered as interference as well. Finally, the outage event of S_1 is calculated as $\mathcal{O}_{D,1}^{(1)}(\mathbf{H}) = \mathcal{E}_D^{(1)}(\mathbf{H}) \cap \mathcal{E}_{D,1}^{(1)}(\mathbf{H})$.

Case 2: The relay cooperates with S_1. The outage event at the destination $\mathcal{E}_D^{(2)}(\mathbf{H})$ is calculated as

$$\mathcal{E}_D^{(2)}(\mathbf{H}) = \mathcal{E}_{D,1|2}^{(2)}(\mathbf{H}) \cup \mathcal{E}_{D,2|1}^{(2)}(\mathbf{H}) \cup \mathcal{E}_{D,1,2}^{(2)}(\mathbf{H}) \qquad [7.13]$$

where

$$\mathcal{E}_{D,1|2}^{(2)}(\mathbf{H}) = \left\{ \alpha I(x_1; \mathbf{y}_D^{(1)} \mid x_2) + \bar{\alpha} I(x_R; \mathbf{y}_D^{(2)}) < r \right\} \qquad [7.14]$$

$$\mathcal{E}_{D,2|1}^{(2)}(\mathbf{H}) = \left\{ \alpha I(x_2; \mathbf{y}_D^{(1)} \mid x_1) < r \right\} \qquad [7.15]$$

$$\mathcal{E}_{D,1,2}^{(2)}(\mathbf{H}) = \left\{ \alpha I(x_1, x_2; \mathbf{y}_D^{(1)}) + \bar{\alpha} I(x_R; \mathbf{y}_D^{(2)}) < 2r \right\} \qquad [7.16]$$

To calculate the outage event of S_1, we define the following event

$$\mathcal{E}_{D,1}^{(2)}(\mathbf{H}) = \left\{ \alpha I(x_1; \mathbf{y}_D^{(1)}) + \bar{\alpha} I(x_R; \mathbf{y}_D^{(2)}) < r \right\} \qquad [7.17]$$

in which the destination treats the signal x_2 as interference during the first transmission phase. Finally, the outage event of S_1 is calculated as $\mathcal{O}_{D,1}^{(2)}(\mathbf{H}) = \mathcal{E}_D^{(2)}(\mathbf{H}) \cap \mathcal{E}_{D,1}^{(2)}(\mathbf{H})$.

Case 3: The relay cooperates with source 2. Swapping the roles of sources 1 and 2, the outage event at the destination $\mathcal{E}_D^{(3)}(\mathbf{H})$ is identical to the previous case. To calculate the outage event of the source S_1, we define the event $\mathcal{E}_{D,1}^{(3)}(\mathbf{H})$ as in [7.12]. Thus, the outage event of S_1 is calculated as $\mathcal{O}_{D,1}^{(3)}(\mathbf{H}) = \mathcal{E}_D^{(3)}(\mathbf{H}) \cap \mathcal{E}_{D,1}^{(3)}(\mathbf{H})$.

Case 4: The relay does not cooperate. The outage at the destination $\mathcal{E}_D^{(4)}(\mathbf{H})$ occurs if the source-to-destination MAC is in outage, which is calculated similar to [7.5]. The outage event of S_1 can also be derived as $\mathcal{O}_{D,1}^{(4)}(\mathbf{H}) = \mathcal{E}_D^{(4)}(\mathbf{H}) \cap \mathcal{E}_{D,1}^{(4)}(\mathbf{H})$, with $\mathcal{E}_{D,1}^{(4)}(\mathbf{H})$ calculated as in [7.12].

Finally, the outage event of S_1 in the error-prone SOMARC/JNCC can be expressed as

$$\mathcal{O}_{D,1}(\mathbf{H}) = \bigcup_{i=1}^{4} \left(\mathcal{Q}_R^{(i)}(\mathbf{H}) \cap \mathcal{O}_{D,1}^{(i)}(\mathbf{H}) \right) \qquad [7.18]$$

The above outage event is conditional on the channel state **H**. The information outage probability for S_1 is then obtained as averaging [7.18] over the fading coefficients

$$P_{out,1} = \int_{\mathbf{H}} [\mathcal{O}_{D,1}(\mathbf{H})] \ p(\mathbf{H}) d(\mathbf{H}) \qquad [7.19]$$

where $p(\mathbf{H})$ is the pdf of **H**.

The ϵ-outage capacity of S_1 is defined as the largest rate of S1 such that its corresponding information outage probability for a given transmission protocol, is smaller than or equal to ϵ.

7.3.2. *Outage analysis of SOMARC/SNCC*

In the case of SNCC/SNCD, we still have two MACs at the relay and destination corresponding to the first time slot, and a point-to-point channel corresponding to the second time slot. The outage event analysis at the relay remains the same as in section 7.3.1. However, the receive signals at the destination from the sources and the relay are now decoded separately. Therefore, the outage events analysis at the destination related to the first time slot exactly follows the one of the relay. More specifically, the outage events $\mathcal{E}_D(\mathbf{H})$, $\mathcal{E}_{D,1}(\mathbf{H})$, and $\mathcal{E}_{D,2}(\mathbf{H})$ are defined similarly to $\mathcal{E}_R(\mathbf{H})$, $\mathcal{E}_{R,1}(\mathbf{H})$, $\mathcal{E}_{R,2}(\mathbf{H})$ (by replacing the subscript R by D). In addition, let $\mathcal{E}_{RD}(\mathbf{H})$ denote the outage event of the relay-to-destination channel:

$$\mathcal{E}_{RD}(\mathbf{H}) = \left\{ \bar{\alpha} I(x_R; \mathbf{y}_D^{(2)}) < r \right\} \qquad [7.20]$$

where r bits per channel use is recalled to be the spectral efficiency of the relay transmission. Depending on the relay transmitted signal, we distinguish four different cases at the destination:

Case 1: The relay cooperates with both sources. If the message of S_1 cannot be correctly decoded from the MAC, it can be recovered, provided that the destination can successfully decode the message of S_2 during the first time slot, and the relay message during the second time slot. Thus, the outage event of S_1 is calculated as

$$\mathcal{O}_{D,1}^{(1)}(\mathbf{H}) = (\mathcal{E}_D(\mathbf{H}) \cap \mathcal{E}_{D,1}(\mathbf{H}) \cap \mathcal{E}_{RD}(\mathbf{H}))$$

$$\cup \left(\mathcal{E}_D(\mathbf{H}) \cap \mathcal{E}_{D,1}(\mathbf{H}) \cap \bar{\mathcal{E}}_{RD}(\mathbf{H}) \cap \mathcal{E}_{D,2}(\mathbf{H})\right) \qquad [7.21]$$

Case 2: The relay cooperates with S_1. The outage event of S_1 is expressed as

$$\mathcal{O}_{D,1}^{(2)}(\mathbf{H}) = (\mathcal{E}_D(\mathbf{H}) \cap \mathcal{E}_{D,1}(\mathbf{H}) \cap \mathcal{E}_{RD}(\mathbf{H})) \qquad [7.22]$$

Case 3: The relay cooperates with S_2. The outage event of S_1 can be expressed as

$$\mathcal{O}_{D,1}^{(3)}(\mathbf{H}) = (\mathcal{E}_D(\mathbf{H}) \cap \mathcal{E}_{D,1}(\mathbf{H})) \qquad [7.23]$$

Case 4: The relay does not cooperate. The outage event of S_1 is calculated as

$$\mathcal{O}_{D,1}^{(4)}(\mathbf{H}) = (\mathcal{E}_D(\mathbf{H}) \cap \mathcal{E}_{D,1}(\mathbf{H})) \qquad [7.24]$$

Finally, the outage event of S_1 in the error-prone SOMARC/SNCC can be expressed as

$$\mathcal{O}_{D,1}(\mathbf{H}) = \bigcup_{i=1}^{4} \left(\mathcal{Q}_R^{(i)}(\mathbf{H}) \cap \mathcal{O}_{D,1}^{(i)}(\mathbf{H})\right) \qquad [7.25]$$

Here again, the outage event $\mathcal{O}_{D,1}(\mathbf{H})$ is conditional on the channel state $\mathbf{H}$, and the information outage probability for S_1, that is $P_{out,1}$, is derived as

$$P_{out,1} = \int_{\mathbf{H}} [\mathcal{O}_{D,1}(\mathbf{H})] \ p(\mathbf{H})d(\mathbf{H}). \qquad [7.26]$$

7.3.3. *Types of input distributions*

7.3.3.1. *Gaussian i.i.d. inputs*

In this case, the mutual information is given by:

$$I(x_i; \mathbf{y}_R^{(1)} \mid x_j) = \log\left(1 + \frac{P_{iR}\,\|\mathbf{h}_{iR}\|^2}{N_0}\right) \qquad [7.27]$$

for $i, j \in \{1, 2\}$, and

$$I(x_1, x_2; \mathbf{y}_R^{(1)}) = \log\det\left(\mathbf{I}_{N_R} + \frac{1}{N_0}\mathbf{H}_R\mathbf{K}_x\mathbf{H}_R^{\dagger}\right) \qquad [7.28]$$

where $\mathbf{H}_R = [\mathbf{h}_{1R}\ \mathbf{h}_{2R}]$ and $\mathbf{K}_x = \operatorname{diag}(P_{1R}, P_{2R})$. The maximum mutual information in [7.8] can also be calculated as

$$I(x_i; \mathbf{y}_R^{(1)}) = \log\left(1 + P_{iR}\mathbf{h}_{iR}^{\dagger}\mathbf{V}_j^{-1}\mathbf{h}_{iR}\right) \qquad [7.29]$$

where $\mathbf{V}_j = N_0\mathbf{I}_{N_R} + P_{jR}\mathbf{h}_{jR}\mathbf{h}_{jR}^{\dagger}$. The other expressions can be derived similarly.

7.3.3.2. *Discrete i.i.d. inputs*

The assumption of Gaussian inputs can be justified only in the case of large signal constellations. In practical systems, the channel inputs are selected from a finite and discrete alphabet (typically QPSK or 16QAM). Thus, to make fair outage comparisons, discrete channel inputs should be chosen from the constellations $\mathcal{X}_i$ of order 2^{q_i}, $i \in \{1, 2, R\}$. Here, we compute the mutual information assuming uniform input distributions. Let x denote the uniformly distributed channel input chosen from the constellation $\mathcal{X}$ of order 2^q, and $\mathbf{y}$ denote the corresponding channel output. The mutual information is derived numerically [UNG 82] as

$$I(x; \mathbf{y}) = q - \mathbb{E}\left[\log_2 \frac{\sum_{\tilde{x} \in \mathcal{X}} p(\mathbf{y}|\tilde{x})}{p(\mathbf{y}|x)}\right] \qquad [7.30]$$

where the expectation is with respect to $p(x, \mathbf{y}) = 2^{-q} p(\mathbf{y}|x)$. All the expressions of the mutual information can be derived similarly in the case of discrete channel inputs.

7.3.4. *Information outage probability achieving codebooks*

To achieve the information outage probability bounds, the codebooks ζ_1, ζ_2, ζ_{1R}, ζ_{2R}, and ζ_{12R} should be universal codebooks. As defined in [TSE 05], a universal codebook of a given rate is a codebook that simultaneously achieves reliable communication over every channel that is not in outage for the chosen rate. Finally, it is worth stressing that, in practice, there exist codebooks with finite lengths whose performance is very close to the ones of universal codebooks. The simulation section (section 7.6) exemplifies such codebook constructions based on convolutional or turbo codes.

7.4. Joint network channel coding and decoding

In this section, we make explicit our proposed JNCC/JNCD approach. We explain the structure of the encoders, when and how JNCC is performed, and the structure of the corresponding multiuser receivers.

7.4.1. *Coding at the sources*

The messages of the two sources are binary vectors $\mathbf{u}_1 \in \mathbb{F}_2^K$ and $\mathbf{u}_2 \in \mathbb{F}_2^K$ of length K. Each source employs a bit-interleaved coded modulation (BICM) [CAI 98]. Binary vectors are first encoded with linear systematic binary encoders C_i : $\mathbb{F}_2^K \rightarrow \mathbb{F}_2^{n_i}$, $i = \{1, 2\}$ into binary codewords $\mathbf{c}_i \in \mathbb{F}_2^{n_i}$ of respective lengths n_i. The codes ζ_i are in general turbo codes, consisting of two recursive systematic convolutional codes (RSCs), denoted by $\mathrm{RSC}_{i,1}$ and $\mathrm{RSC}_{i,2}$, concatenated in parallel using optimized semi-random interleavers $\pi_{0,i}$. The coded bits are then interleaved using interleavers Π_i and reshaped as two binary matrices $\mathbf{V}_i \in \mathbb{F}_2^{\alpha N \times q_i}$. Memoryless modulators based on one-to-one binary labeling maps $\phi_i : \mathbb{F}_2^{q_i} \rightarrow \mathscr{X}_i$ transform the binary arrays $\mathbf{V}_i$ into the complex vectors $\mathbf{x}_i \in \mathscr{X}_i^{\alpha N}$. For ϕ_i, we choose Gray labeling. In the following, we denote by $v_{i,k,\ell} = \phi_{i,\ell}^{-1}(x_{i,k})$ the ℓth bit of the binary labeling of each symbol $x_{i,k}$ for $i \in \{1, 2\}$ and $k = 1, \cdots, \alpha N$.

7.4.2. *Relaying function*

Relay processing is divided into two steps, as shown in Figure 7.2. During the first time slot, based on [7.1], the relay performs a joint detection and decoding procedure to obtain the hard binary estimation of the information bits, $\hat{\mathbf{u}}_i \in \mathbb{F}_2^K$. On the basis of this estimation, the relay chooses an SDF approach for cooperation. Different cases can then be distinguished, depending on the number of successfully decoded messages. In the following, first, we briefly describe the relay detection and decoding algorithm, and then we detail our proposed JNCC scheme.

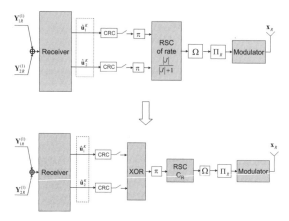

Figure 7.2. *Block diagram of the sequential processing at relay*

7.4.2.1. *Relay detection and decoding*

The joint detection and decoding is performed in a suboptimal iterative way [HAG 97]. An inner soft-in soft-out (SISO) maximum *a posteriori* (MAP) detector generates extrinsic information on coded bits using the received signal [7.1] and *a priori* information coming from the outer SISO decoders SISO$_1$ and SISO$_2$ (referring to the decoding of ζ_1 and ζ_2). For the general case of turbo codes at the sources, the outer SISO decoder of S_i generates extrinsic information on both systematic and coded bits of S_i by activating the SISO decoder SISO$_{i,1}$ corresponding to RSC$_{i,1}$, and then SISO$_{i,2}$ corresponding to RSC$_{i,2}$. It is important to remember that each SISO decoding stage takes into account all the available a priori information on systematic bits [RAP 98] (and Algorithm 1 of section 7.4.3.2). The extrinsic information on both source codewords is then interleaved and fed back to the detector, which in turn employs it as *a priori* information for the next iteration. It is worth noting that the proper (de)multiplexing and (de)puncturing are also performed if needed. The process is repeated until convergence. For the representation of the input/output soft information, we use log ratios of probabilities. The log *a posteriori* probability ratio (LAPPR) on bit $v_{i,k,\ell} = \phi_{i,\ell}^{-1}(x_{i,k})$ delivered by the SISO MAP detector is defined as:

$$\Lambda(v_{i,k,\ell}) = \log \frac{P(v_{i,k,\ell} = 1|\mathbf{y}_{R,k}^{(1)})}{P(v_{i,k,\ell} = 0|\mathbf{y}_{R,k}^{(1)})} \qquad [7.31]$$

and, in practice, evaluated as

$$\Lambda(v_{i,k,\ell}) \simeq \log \frac{\sum_{a\in\mathscr{X}_i : \phi_{i,\ell}^{-1}(a)=1, b\in\mathscr{X}_j} P(\mathbf{y}_{R,k}^{(1)}|x_{i,k} = a, x_{j,k} = b)e^{\xi(a)+\xi(b)}}{\sum_{a\in\mathscr{X}_i : \phi_{i,\ell}^{-1}(a)=0, b\in\mathscr{X}_j} P(\mathbf{y}_{R,k}^{(1)}|x_{i,k} = a, x_{j,k} = b)e^{\xi(a)+\xi(b)}}$$

$$[7.32]$$

for $i, j \in \{1, 2\}, i \neq j$, with,

$$\xi(a) = \sum_{\ell'=1}^{\log_2 |\mathscr{X}_i|} \phi_{i,\ell'}^{-1}(a) E(v_{i,k,\ell'}) \qquad [7.33]$$

$$\xi(b) = \sum_{\ell'=1}^{\log_2 |\mathscr{X}_j|} \phi_{j,\ell'}^{-1}(b) E(v_{j,k,\ell'}) \qquad [7.34]$$

where $\{E(v_{i,k,\ell})\}$ and $\{E(v_{j,k,\ell})\}$ are log *a priori* probability ratios (LAPRs) on bits $v_{i,k,\ell}$ and $v_{j,k,\ell}$ provided by the SISO decoders SISO$_1$ and SISO$_2$. The extrinsic information on bit $v_{i,k,\ell}$ is given by $L(v_{i,k,\ell}) = \Lambda(v_{i,k,\ell}) - E(v_{i,k,\ell})$, and, after deinterleaving, feeds the corresponding outer SISO decoder.

7.4.2.2. JNCC

As previously mentioned, the relay chooses an SDF approach for cooperation, which is based on the number of successfully decoded messages, the knowledge of which being ensured by using cyclic redundancy check (CRC) codes for each source message. Let $J = \{j_1, \ldots, j_{|J|}\}$, $|J| \leq 2$ denote the set of message indices that have been successfully decoded. For the case where $J = \emptyset$, the relay does not cooperate. Otherwise, it interleaves each message $\mathbf{u}_j$, $j \in J$, by π. The interleaved binary streams are then linearly combined over $\mathbb{F}_2$ using binary linear encoders

$$\begin{cases} C_{R,1} : \mathbb{F}_2^K \to \mathbb{F}_2^{2K} \\ C_{R,2} : \mathbb{F}_2^K \times \mathbb{F}_2^K \to \mathbb{F}_2^{3K} \end{cases} \qquad [7.35]$$

for $J \neq \emptyset$. For $C_{R,|J|}$, we choose an RSC encoder of rate $|J|/|J| + 1$, defined by the generator matrix

$$G_{R,|J|}(D) = \left[\begin{array}{c|c} & \frac{p_1(D)}{q(D)} \\ I_{|J|} & \vdots \\ & \frac{p_{|J|}(D)}{q(D)} \end{array} \right] \qquad [7.36]$$

where $p_i(D)$, $i \in \{1, ..., |J|\}$, and $q(D)$ are the generator polynomials of the encoder $C_{R,|J|}$. This yields the binary vector $\mathbf{c}_R \in \mathbb{F}_2^{K(|J|+1)}$. A linear transformation $\Omega : \mathbb{F}_2^{K(|J|+1)} \to \mathbb{F}_2^K$ is then applied, which selects only the parity bit of $\mathbf{c}_R$ to obtain the new vector $\mathbf{c}'_R \in \mathbb{F}_2^K$. As all the systematic bits are removed, this structure maximizes the spectral efficiency. The vector $\mathbf{c}'_R$ is bit interleaved using the interleaver Π_R and reshaped as a binary matrix $\mathbf{V}_R \in \mathbb{F}_2^{\bar{\alpha}N \times q_R}$. Finally, a memoryless modulator based on a one-to-one binary labeling map $\phi_R : \mathbb{F}_2^{q_R} \to \mathscr{X}_R$ transforms the binary array $\mathbf{V}_R$ into the complex vector $\mathbf{x}_R \in \mathscr{X}_R^{\bar{\alpha}N}$. $\mathscr{X}_R \subset \mathbb{C}$ is a complex constellation of order $|\mathscr{X}_R| = 2^{q_R}$ with energy normalized to unity. For ϕ_R, we choose Gray labeling. In the following, we denote by $v_{R,k,\ell} = \phi_{R,\ell}^{-1}(x_{R,k})$ the ℓth bit of the binary labeling of each symbol $x_{R,k}$ for $k = 1, \cdots, \bar{\alpha}N$. Finally, to let the destination detect which of the messages are included in the relay signal, the relay transmits side information (one additional bits) to indicate its state to the receiver.

The proposed CODING scheme entails the need of a decoder corresponding to the code of rate $|J|/|J|+1$, which increases the complexity while $|J|$ becomes larger. Furthermore, it does not guarantee the full diversity for low memory orders, as shown in [HAT 11]. But if we assume that all the feedforward generators of $G_{R,|J|}(D)$ are the same, that is $p_i(D) = p(D)$, we obtain an equivalent model, as depicted in Figure 7.2, which ensures full diversity and simplifies the decoder structure. In this model, the relay combines all the correctly decoded messages by XOR, that is $\mathbf{u}_R = \mathbf{u}_{j_1} \oplus \cdots \oplus \mathbf{u}_{j_{|J|}}$. The resulting vector is then interleaved by π, and encoded to $\mathbf{c}_R$ using a binary linear encoder $C_R : \mathbb{F}_2^K \to \mathbb{F}_2^{2K}$. For C_R, we choose the RSC encoder defined by the generator matrix $G_R(D) = \begin{bmatrix} 1 & p(D)/q(D) \end{bmatrix}$. The selection function Ω then removes the systematic bits. The rest of the operations remain the same. The XOR operation ensures full diversity for the OMARC using SNCC [WOL 07] or JNCC [DUY 10]. As shown in Appendix A, the high SNR slope of the outage probability of MAC versus SNR (in dB scale), for the critical case of just one receive antenna is the same as the one of the orthogonal MAC. Thus, the full diversity design for OMARC remains valid when we have collisions at the relay and destination.

7.4.3. *JNCD at the Destination*

The JNCD at the destination depends on the side information received from the relay: In case 1, where the relay has successfully decoded both source messages, two distributed turbo codes are formed at the destination. In case 2 (case 3), where the relay has successfully decoded the information of source 1 (source 2), one distributed turbo code is formed at the destination corresponding to source 1 (source 2), and a separate decoder corresponding to C_2 (C_1) is used to decode the information of the other source. In these cases, at the end of the second transmission time slot, the destination starts to detect and decode the original data, processing the received signals [7.2] and [7.3]. Finally, in case 4, where the relay does not cooperate, the destination applies iterative detection and decoding, processing the received signal [7.2], and using the two separate decoders corresponding to C_1 and C_2. Here again, we resort to a suboptimal iterative procedure. Extrinsic information on coded bits circulates between SISO MAP detector and demapper corresponding to two transmission time slots and the outer decoders, while, at the same time, extrinsic information on systematic bits circulates between the SISO decoders of each code.

7.4.3.1. *SISO MAP detector and demapper*

The SISO MAP detector computes the LAPPR $\Lambda(v_{i,k,\ell})$ with $v_{i,k,\ell} = \phi_{i,\ell}^{-1}(x_{i,k})$, $i \in \{1, 2\}$, using the received signal [7.2] and *a priori* information coming from the outer SISO decoders. Expression is similar to [7.32] substituting $\mathbf{y}_{D,k}^{(1)}$ for $\mathbf{y}_{R,k}^{(1)}$. We now turn to the SISO MAP demapper, which delivers soft information on the additional relay parity bits in case of relay cooperation (successful selective relaying). The LAPPR on bit $v_{R,k,\ell} = \phi_{R,\ell}^{-1}(x_{R,k})$ is defined as

$$\Lambda(v_{R,k,\ell}) = \log \frac{P(v_{R,k,\ell} = 1 | \mathbf{y}_{D,k}^{(2)})}{P(v_{R,k,\ell} = 0 | \mathbf{y}_{D,k}^{(2)})} \qquad [7.37]$$

and evaluated as

$$\Lambda(v_{R,k,\ell}) \simeq \log \frac{\sum_{c \in \mathscr{X}_R : \phi_{R,\ell}^{-1}(c)=1} P(\mathbf{y}_{D,k}^{(2)} | x_{R,k} = c) e^{\xi(c)}}{\sum_{c \in \mathscr{X}_R : \phi_{R,\ell}^{-1}(c)=0} P(\mathbf{y}_{D,k}^{(2)} | x_{R,k} = c) e^{\xi(c)}} \qquad [7.38]$$

with

$$\xi(c) = \sum_{\ell'=1}^{\log_2 |\mathscr{X}_R|} \phi_{R,\ell'}^{-1}(c) E(v_{R,k,\ell'}) \qquad [7.39]$$

where $\{E(v_{R,k,\ell})\}$ is the LAPR on bit $v_{R,k,\ell}$ provided by the SISO decoder SISO$_R$ corresponding to the relay joint network-channel encoder ($C_{R,|J|}$ or XOR followed by C_R). Finally, the extrinsic information on $v_{R,k,\ell}$ is given by $L(v_{R,k,\ell}) = \Lambda(v_{R,k,\ell}) - E(v_{R,k,\ell})$, and, after de-interleaving, feeds SISO$_R$.

7.4.3.2. *Message-passing schedule*

A recapitulative block diagram of the JNCD is depicted in Figure 7.3. In this section, we detail the message-passing for the case where the relay cooperates with both the sources using a XOR followed by a linear encoding. We also consider the case of turbo codes at the sources, that is each C_i, $i \in \{1,2\}$ consists of two RSC encoders separated by $\pi_{0,i}$. The generalization to other cases is straightforward. The SISO decoder SISO$_i$ corresponds to C_i, $i \in \{1,2\}$, and SISO$_R$

corresponds to the relay encoder (XOR followed by C_R). SISO decoder j refers to the decoding of C_j, $j \in \{1, 2, R\}$. Each SISO$_i$, $i \in \{1, 2\}$, is made up of the two SISO decoders SISO$_{i,1}$ and SISO$_{i,2}$. Let $\mathbf{L}_{s_i}$, $\mathbf{L}_{p_i}$, and $\mathbf{L}_{p_R}$, $i \in \{1, 2\}$, denote, respectively, the soft information of the systematic and parity bits of the two sources and the relay, obtained from the channel MAP detector and demapper. It is worth noting that the proper (de)multiplexing and (de)puncturing are also performed if needed. In Figure 7.3, the (de)puncturing is included in the blocks corresponding to (de)multiplexing. Let $\mathbf{E}_{s_i(j)}$, $\mathbf{E}_{p_i(j)}$, and $\mathbf{E}_{p_R(j)}$ denote the extrinsic information generated by SISO$_j$, $j \in \{1, 2, R\}$. Similarly, let $\mathbf{L}_{p_{i,1}}$ and $\mathbf{L}_{p_{i,2}}$ denote, respectively, the soft information of the parity bits corresponding to SISO$_{i,1}$ and SISO$_{i,2}$ obtained from the MAP detector; $\mathbf{E}_{s_i(i,1)}$ and $\mathbf{E}_{s_i(i,2)}$ denote, respectively, the extrinsic information on systematic bits generated by SISO$_{i,1}$ and SISO$_{i,2}$; and $\mathbf{E}_{p_i(i,1)}$ and $\mathbf{E}_{p_i(i,2)}$ denote, respectively, the extrinsic information on parity bits generated by SISO$_{i,1}$ and SISO$_{i,2}$.

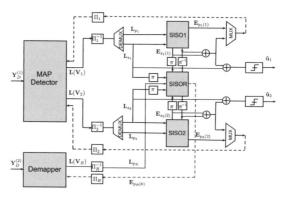

Figure 7.3. *JNCD at the destination (relay cooperates with both sources)*

The SISO MAP detector generates the LAPPRs for the systematic and parity bits in $\mathbf{V}_1$ using $\mathbf{E}_{s_1(1)} + \pi^{-1}(\mathbf{E}_{s_1(R)})$ and

$\mathbf{E}_{p_1(1)}$, respectively (after proper multiplexing interleaving). It also generates the LAPPRs for the systematic and parity bits in $\mathbf{V}_2$ using $\mathbf{E}_{s_2(2)} + \pi^{-1}(\mathbf{E}_{s_2(R)})$ and $\mathbf{E}_{p_2(2)}$, respectively. It is worth stressing that $\mathbf{E}_{s_1(1)} = \mathbf{E}_{s_1(1,1)} + \pi_{0,1}^{-1}(\mathbf{E}_{s_1(1,2)})$, and $\mathbf{E}_{s_2(2)} = \mathbf{E}_{s_2(2,1)} + \pi_{0,2}^{-1}(\mathbf{E}_{s_2(2,2)})$, as depicted in Figure 7.4. The MAP demapper generates the LAPPRs for the parity bits in $\mathbf{V}_R$ using $\mathbf{E}_{p_R(R)}$. Then, the two distributed decoders are activated and the extrinsic information for both the systematic and parity bits calculated, which are fed back to the SISO MAP detector and demapper.

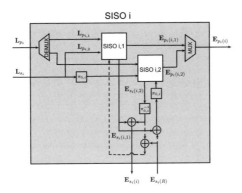

Figure 7.4. *SISO decoder i in case of compound codes at sources*

In the case of an XOR encoding scheme (full diversity by construction), we detail in Figure 7.5 and hereafter the low complexity implementation of SISO$_R$. As depicted in Figure 7.5, the SISO decoder corresponding to C_R (DEC$_R$) should collect all the *a priori* information $\mathbf{L}_{u_R}$ on $\mathbf{u}_R$. Denoting $\mathbf{L}_1 = \pi(\mathbf{L}_{s_1} + \mathbf{E}_{s_1(1)})$ and $\mathbf{L}_2 = \pi(\mathbf{L}_{s_2} + \mathbf{E}_{s_2(2)})$, it yields, taking into account the XOR constraint node (see, e.g. [HAG 96]),

$$L_{u_R,k} = \log \frac{e^{L_{1,k}} + e^{L_{2,k}}}{1 + e^{(L_{1,k}+L_{2,k})}}. \qquad [7.40]$$

As well known in the framework of the LDPC sum product algorithm, any output message (or extrinsic information) originating from an XOR constraint node can be obtained from the two other input messages following [7.40]. Note that independency between messages should hold in order to apply [7.40]. Finally, SISO_R computes at its output, the extrinsic information $\mathbf{E}_{s_i(R)}$ from $\mathbf{L}_j$ and $\mathbf{E}_{u_R(R)}$, $i, j \in \{1, 2\}$, $i \neq j$, where $\mathbf{E}_{u_R(R)}$ is the extrinsic information on u_R computed by the decoder corresponding to C_R. The message-passing schedule for the JNCD at each iteration and the final hard decisions are recapitulated in Algorithm 1.

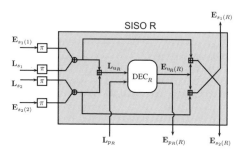

Figure 7.5. *XOR decoder*

7.5. Separate network channel coding and decoding

As previously mentioned, the SNCC scheme is based on the XOR operation at the relay, and the network-coded signal is separately decoded at the destination. Thus, in case of SOMARC/SNCC, a joint detection and decoding procedure similar to section 7.4.2.1 is performed at the destination on the signal received during the first transmission time slot, and a separate decoding is performed on the relay signal received during the second time slot. The channel decoders then make hard decisions and output the estimates to the network decoder. If at least two out of three channel output estimates are error-free, the network decoder can retrieve both source messages.

Algorithm 1: JNCD at the destination

(INITIALIZATION)

Set all the a priori information to zero.

(ITERATIONS)

Iterate until convergence:

1) Activate the SISO MAP detector using the received signal $\mathbf{Y}_D^{(1)}$, and the messages $\mathbf{E}_{s_1(1)} + \pi^{-1}(\mathbf{E}_{s_1(R)})$, $\mathbf{E}_{p_1(1)}$ and $\mathbf{E}_{s_2(2)} + \pi^{-1}(\mathbf{E}_{s_2(R)})$, $\mathbf{E}_{p_2(2)}$, where $\mathbf{E}_{s_i(i)} = \mathbf{E}_{s_i(i,1)} + \pi_{0,i}^{-1}(\mathbf{E}_{s_i(i,2)})$.

2) Activate the SISO MAP demapper using the received signal $\mathbf{Y}_D^{(2)}$, and the message $\mathbf{E}_{p_R(R)}$.

3) Activate simultaneously the SISO decoders SISO_1 and SISO_2

   i) Activate simultaneously the $\text{SISO}_{1,1}$ and $\text{SISO}_{2,1}$ with the messages $\mathbf{L}_{s_1}$, $\mathbf{L}_{p_{1,1}}$ and $\mathbf{L}_{s_2}$, $\mathbf{L}_{p_{2,1}}$ provided by the MAP detector, and $\pi_{0,1}^{-1}(\mathbf{E}_{s_1(1,2)}) + \pi^{-1}(\mathbf{E}_{s_1(R)})$ and $\pi_{0,2}^{-1}(\mathbf{E}_{s_2(2,2)}) + \pi^{-1}(\mathbf{E}_{s_2(R)})$, which are derived from the previous iteration.

   ii) Activate simultaneously the $\text{SISO}_{1,2}$ and $\text{SISO}_{2,2}$ with, respectively, the messages $\pi_{0,1}(\mathbf{L}_{s_1})$, $\mathbf{L}_{p_{1,2}}$ and $\pi_{0,2}(\mathbf{L}_{s_2})$, $\mathbf{L}_{p_{2,2}}$ provided by the MAP detector, and $\pi_{0,1}(\mathbf{E}_{s_1(1,1)}) + \pi_{0,1} \circ \pi^{-1}(\mathbf{E}_{s_1(R)})$ and $\pi_{0,2}(\mathbf{E}_{s_2(2,1)}) + \pi_{0,2} \circ \pi^{-1}(\mathbf{E}_{s_2(R)})$.

4) Activate the SISO decoder SISO_R with the messages $\mathbf{L}_{p_R}$ provided by the MAP demapper, and $\mathbf{L}_1 = \pi(\mathbf{L}_{s_1} + \mathbf{E}_{s_1(1)})$ and $\mathbf{L}_2 = \pi(\mathbf{L}_{s_2} + \mathbf{E}_{s_2(2)})$.

(HARD DECISIONS)

Combine all the available information on the systematic bits $\mathbf{u}_1$ and $\mathbf{u}_2$:

$$\mathbf{L}_{s_1} + \mathbf{E}_{s_1(1,1)} + \pi_{0,1}^{-1}(\mathbf{E}_{s_1(1,2)}) + \pi^{-1}(\mathbf{E}_{s_1(R)}) \to \hat{\mathbf{u}}_1$$

$$\mathbf{L}_{s_2} + \mathbf{E}_{s_2(2,1)} + \pi_{0,2}^{-1}(\mathbf{E}_{s_2(2,2)}) + \pi^{-1}(\mathbf{E}_{s_2(R)}) \to \hat{\mathbf{u}}_2$$

7.6. Numerical results

In this section, we provide some numerical results to evaluate the effectiveness of our approach. In our comparisons, we consider both SOMARC and OMARC using JNCC or SNCC. We start by detailing the topology of the network. For the sake of simplicity, we consider a symmetric MARC, that is $d_{1R} = d_{2R}$ and $d_{1D} = d_{2D}$. The average energy per available dimension allocated to the two sources is the same, that is $P_{0,1} = P_{0,2} = P_0$. We fix the same path loss factor, that is $\kappa = 3$, free distance, that is $d_0 = 1$, and noise power spectral density, that is $N_0 = 1$, for all links. Due to the half-duplex nature of the relay, the transmission time slot of the sources and the relay are separated in time. We fix $\alpha = 2/3$ for SOMARC, which yields $P_1 = P_2 = 3/2P_0$. In OMARC, the two sources transmit in consecutive, equal duration, time slots. Thus, the first two time slots are dedicated to the sources, and the third to the relay. It arrives that $P_1 = P_2 = 3P_0$ for OMARC. The relay, in case of cooperation, always transmits at $P_R = 3P_{0,R}$ for both OMARC and SOMARC. For simulation purposes, two different configurations are considered: In the first configuration, we fix the number of receive antennas to one both at the relay and destination, that is $N_R = N_D = 1$. The geometry is chosen such that $d_{ij} = d_{RD} = d$, which yields $P_{i,j} = P_{RD} = \gamma$ for SOMARC and $P_{i,j} = 2\gamma$, $P_{RD} = \gamma$ for OMARC, $i \in \{1, 2\}$, $j \in \{R, D\}$, where γ is the receive SNR per symbol or dimension. In the second configuration, we increase the number of receive antennas at the destination to 4, that is $N_R = 1$ and $N_D = 4$. The geometry is chosen such that $d_{iR} = d_1$ and $d_{iD} = d_{RD} = d$ with $(d_1/d)^{-3} = 100$, $i \in \{1, 2\}$. This yields $P_{iR} = 100\gamma$ (or $\gamma + 20$ in dB) and $P_{iD} = P_{RD} = \gamma$ for SOMARC, which translates into $P_{iR} = 200\gamma$, $P_{iD} = 2\gamma$, and $P_{RD} = \gamma$ for OMARC, $i \in \{1, 2\}$. Each message of the sources has length $K = 1,024$ information bits. In our proposed distributed

JNCC, the complex signal sets $\mathscr{X}_1$, $\mathscr{X}_2$, and $\mathscr{X}_R$ used in BICM are either QPSK or 16QAM constellation (Gray labeling) and their corresponding sum rates are $\eta = 4/3$ bits per channel use (b./c.u) and $\eta = 8/3$ b./c.u, respectively.

7.6.1. *Information-theoretic comparison of the protocols*

7.6.1.1. *Individual ϵ-outage capacity with Gaussian inputs*

In the first set of simulations, we consider the ϵ-outage capacity of S_1, and we compare the individual ϵ-outage capacity $C_\epsilon(\gamma)$ of JNCC and SNCC for the SOMARC and the OMARC. In our analysis, we fix $\epsilon = 10^{-2}$. The number of receive antennas at the destination is either $N_D = 1$ or $N_D = 4$. The corresponding results are depicted in Figure 7.6. As we can see, the ϵ-outage capacity for the SOMARC is always higher than the ϵ-outage capacity for the OMARC regardless of the network channel coding strategy (i.e. JNCC or SNCC); especially, in the case of $N_D = 4$, JNCC with orthogonal multiple access (OMARC/JNCC) is strictly suboptimal and the ϵ-outage capacity gain of SOMARC/JNCC versus OMARC/JNCC for individual rates above 2b./c.u. is more than 5 dB. This results from the fact that, in the presence of multiple receive antennas, a non-orthogonal MAC can better exploit the available degrees of freedom. Moreover, even in the case of $N_D = 1$, which is not *a priori* favorable for an MAC, we see that SOMARC/JNCC can provide an ϵ-outage capacity gain of approximately 4 dB for data rates above 2 b./c.u. Finally, the JNCC schemes outperform the SNCC ones for both transmission protocols. For the data rate of 2 b./c.u, the ϵ-outage capacity gains are about 5 dB in case of SOMARC for both $N_D = 1$ and $N_D = 4$, 3 dB and 4 dB in case of OMARC with, respectively, $N_D = 1$ and $N_D = 4$.

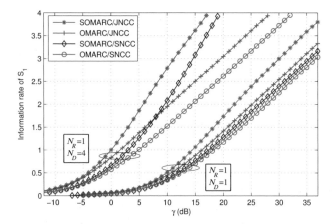

Figure 7.6. *Individual ε-outage capacity – ε = 10⁻² – SOMARC versus OMARC – JNCC versus SNCC*

7.6.1.2. *Individual information outage probability with discrete inputs*

In the second set of simulations, our purpose is first to compare the individual outage probability of SOMARC/JNCC versus OMARC/JNCC, and for the fixed sum rates of $\eta = 4/3$ and $\eta = 8/3$ b./c.u. To achieve the same spectral efficiency as the SOMARC, we consider two approaches for OMARC: (1) we impose on the transmitters to use the same input alphabet as in the case of SOMARC, which makes sense if we want to preserve the same level of peak-to-average power ratio (PAPR); (2) we employ constellation expansion for the sources in OMARC. In the first approach, the two sources have no other choice but to transmit their information symbols without any coding, and thus from a theoretical perspective $(N \to \infty)$, the system is always in outage. In the second approach, the sources increase the cardinality of their modulation while preserving the same spectral efficiency, which makes room for coding. Thus, the performance of SOMARC with QPSK is compared with the performance of OMARC with 16QAM at the sources and QPSK at the relay. Thus, the information

outage probability of SOMARC with QPSK is compared with
the information outage probability of OMARC with 16QAM at
the sources and QPSK at the relay. Similarly, the information
outage probability of SOMARC with 16QAM is compared with
the information outage probability of OMARC with 64QAM
at the sources and 16QAM at the relay. The corresponding
results are depicted in Figure 7.7 for the sum rate of $\eta = 4/3$
b./c.u. and in Figure 7.8 for the sum rate of $\eta = 8/3$ b./c.u.,
for both $N_D = 1$ and $N_D = 4$. As we can see, in all cases, the
information outage probability of SOMARC is smaller than
that of OMARC. Considering the second approach, for $\eta = 8/3$
b./c.u., and at BLER of 10^{-2}, the power gain is approximately
equal to 2.5 dB for $N_D = 1$ and becomes even larger for
$N_D = 4$, attaining 3.5 dB at BLER of 10^{-2}, which reconfirms
the suboptimality of the orthogonal multiple access in case of
multiple receive antennas.

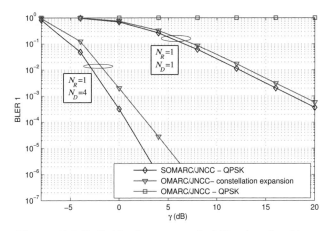

Figure 7.7. *Individual outage probability (e.g. for S_1) –
SOMARC/JNCC versus OMARC/JNCC $\eta = 4/3$ b./c.u.*

To pursue our analysis, we compare the individual
information outage probabilities of SOMARC/JNCC and

SOMARC/SNCC. Here again, to keep the same spectral efficiency for the SNCC case, we have the aforementioned two approaches. Using the first approach, the relay-to-destination channel is always in outage in the case of SOMARC/SNCC, and thus it leads to the performance of an MAC corresponding to the first transmission time slot. This explains the difference of slopes between the two curves in the corresponding figures. In the second approach, constellation expansion is employed for the relay-to-destination channel. Thus, in SOMARC/SNCC, the relay uses 16QAM for $\eta = 4/3$ b./c.u., and 64QAM for $\eta = 8/3$ b./c.u. The corresponding results are depicted in Figures 7.9 and 7.10 for both $N_D = 1$ and $N_D = 4$. As we see, the SOMARC/SNCC has always a performance loss compared to the SOMARC/JNCC. In the case of constellation expansion and $N_D = 1$, at BLER of 10^{-2}, the loss is around 2 dB for $\eta = 4/3$ b./c.u., and 3 dB for $\eta = 8/3$ b./c.u. The performance loss is much higher when we consider $N_D = 4$, and it attains 3 dB for $\eta = 4/3$ b./c.u., and 4 dB for $\eta = 8/3$ b./c.u.

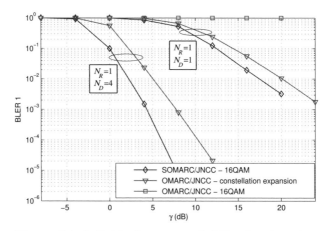

Figure 7.8. *Individual outage probability (e.g. for S_1) – SOMARC/JNCC versus OMARC/JNCC – $\eta = 8/3$ b./c.u.*

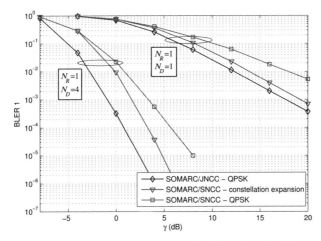

Figure 7.9. *Individual outage probability (e.g. for S_1) – SOMARC / JNCC versus SOMARC / SNCC – $\eta = 4/3$ b. / c.u.*

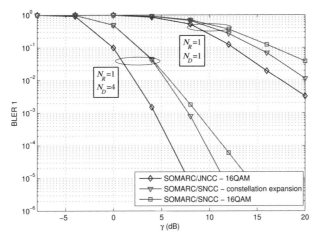

Figure 7.10. *Individual outage probability (e.g. for S_1) – SOMARC / JNCC versus SOMARC / SNCC – $\eta = 8/3$ b. / c.u.*

7.6.2. *Performance of practical code design*

In the following, the number of iterations I is set to 5 at the relay and to 10 (for $N_D = 1$) or 3 (for $N_D = 4$) at the destination. These numbers of iterations ensure convergence and allow us to very closely approach the performance of a Genie Aided (GA) receiver at sufficiently high SNR for the selected modulation and coding schemes, the Genie Aided (GA) receiver corresponding to the ideal case where the interference is known and perfectly removed.

7.6.2.1. *Comparison of JNCC functions: XOR versus general scheme*

In this section, we compare the performance of the two JNCC functions introduced in section 7.4.2.2 for SOMARC. The first one is based on $G_{R,|J|}$, and the second is based on XOR. This experience is carried out with $N_R = N_D = 1$ and with $N_R = 1$ and $N_D = 4$, and for the sum rate of $\eta = 4/3$ b./c.u. As we are interested in comparing the JNCC functions, we assume that the source-to-relay links are error-free in this set of simulations. In our comparisons, we also consider two different coding schemes at the sources: (1) the two sources use identical turbo codes of rate-1/2 made of two 4-state rate-1/2 RSC encoders with generator matrix $G_1 = [\ 1\quad 7/5\ ]$ in octal representation, whose half of the parity bits are punctured; (2) the two sources use identical 64-state rate-1/2 RSC encoder with generator matrix $G_2 = [\ 1\quad 175/133\ ]$. Exhaustive simulations showed that those numbers of states yield the best performance/complexity trade-off. In both of the above schemes, the relay employs 4-state, 16-state, or 64-state RSC encoders. These RSC encoders are represented as $[\ 1\quad p/q\ ]$ and $\begin{bmatrix} 1 & 0 & p_1/q \\ 0 & 1 & p_2/q \end{bmatrix}$ for respective cases of XOR and general scheme, and in octal notation. In the case of a 4-state RSC encoder, $p = 7$ and $q = 5$ for the XOR scheme, and $p_1 = 7$, $p_2 = 2$, and $q = 3$ for the general scheme. In the case of a 16-state RSC encoder, $p = 35$ and $q = 23$ for the XOR scheme, and $p_1 = 27$, $p_2 = 33$, and $q = 31$ for the general scheme. In the case of a 64-state RSC encoder, $p = 175$ and $q = 233$ for the XOR scheme, and $p_1 = 52$, $p_2 = 36$, and $q = 115$ for the general scheme. The SISO decoders implement the BCJR algorithm

[BAH 74]. The system performance is measured in terms of a joint BLER, which is defined as the probability of having at least one erroneously decoded information bit in either of the decoded blocks at the destination.

The simulation results for the first scheme (with turbo codes at sources) are depicted in Figure 7.11. As we have seen that the performance of JNCC based on XOR is not affected by the memory order of the RSC encoder at the relay, only the performance results for the memory order of 4 are depicted. As we see, the JNCC scheme based on XOR achieves the promised full diversity by construction, whatever the memory order of the RSC encoder. The general JNCC scheme also becomes full diversity for a sufficient memory order, but is not as good as the XOR scheme, especially in the case of $N_D = 1$.

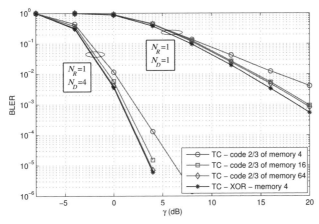

Figure 7.11. *Joint BLER performance – error-free S-R links – Turbo Code (TC) at sources – JNCC based on XOR versus JNCC based on double input binary linear code – $\eta = 4/3$ b./c.u.*

Now, we analyze the performance of the second scheme (with RSC encoders at sources). The simulation results are depicted in Figure 7.12. Since the JNCC based on XOR performs slightly better when the memory order of the

RSC encoder increases, only the performance results for the memory order of 64 are depicted. As we see, here again, the JNCC scheme based on XOR achieves full diversity irrespective of the memory order of the RSC encoder code. The general JNCC scheme also becomes full diversity for a sufficient memory order and would exhibit better coding gain for less severe fading distribution (e.g. with receive antenna diversity) at the expense of more complex decoding. For comparison purposes, we have also plotted the best choice of the first scheme, which is the JNCC based on XOR with turbo codes and RSC encoder of memory order 4 at, respectively, the sources and the relay. Simulation results show that this scheme exhibits the best performance.

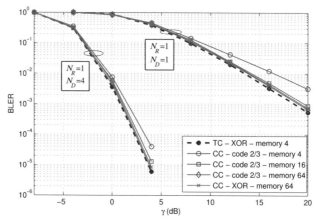

Figure 7.12. *Joint BLER performance – error-free S-R links – Convolutional Code (CC) and Turbo Code (TC) at sources – JNCC based on XOR versus JNCC based on double input binary linear code –* $\eta = 4/3$ *b. /c.u.*

7.6.2.2. *Gap to outage limits*

Here, we first evaluate the gap between the individual BLER of practical designs for SOMARC/JNCC and that of their corresponding information outage probability. The experiment is carried out for $\eta = 4/3$ b./c.u. and with the best coding schemes analyzed in section 7.6.2.1 for both cases of turbo codes and RSC encoders at the sources. Thus, in the first case, the sources use punctured turbo codes made of 4-state RSC encoders, and the relay uses JNCC based on XOR and an RSC encoder of memory order 4. In the second case, the sources use 64-state RSC encoders, and the relay uses JNCC based on the double input 64-state RSC encoder. The corresponding results are demonstrated in Figure 7.13. As expected, the JNCC scheme based on turbo codes provides the best results and performs 1 dB and 1.5 dB away from the information outage probability for respective cases of $N_D = 1$ and $N_D = 4$.

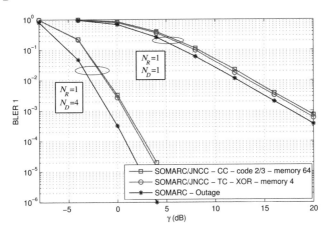

Figure 7.13. *Individual BLER (e.g. for S_1) – Practical SOMARC/JNCC versus outage limit – $\eta = 4/3$ b./c.u.*

Next, we evaluate the individual BLER of practical designs for SOMARC/SNCC versus their corresponding information outage probability. The experiment is carried out for $\eta = 4/3$ b./c.u. and for both cases of turbo codes and RSC encoders at both sources and the relay. In the first case, the sources and the relay use the punctured turbo codes made of 4-state RSC encoders of rate $1/2$ with generator matrix $\mathbf{G}_1$. In the second case, the sources and the relay use RSC encoders of rate $1/2$

with different memory orders of 4, 16, and 64. For the memory order of 4, we use the RSC encoder with generator matrix G_1. For the memory order of 16, we use the RSC encoder with generator matrix $G_3 = [\ 1\ \ 35/23\ ]$, and for the memory order of 64, we use the RSC encoder with generator matrix G_2. The constellation expansion is also performed at the relay. The simulation results are plotted in Figure 7.14 for both $N_D = 1$ and $N_D = 4$. As we see, increasing the memory order of the RSC encoder improves the performance of the second case. But the turbo code remains the best choice which performs 1 dB away from the individual outage probability for both $N_D = 1$ and $N_D = 4$.

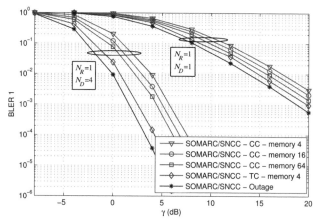

Figure 7.14. *Individual BLER (e.g. for S_1) – practical SOMARC / SNCC versus outage limit – $\eta = 4/3\ b./c.u.$ – constellation expansion at the relay*

7.6.2.3. *Comparison of the different protocols*

In this section, we first compare the individual BLER of practical code design for SOMARC/JNCC with that of the OMARC/JNCC. The JNCC in both protocols is based on XOR. In SOMARC/JNCC, the two sources use the punctured turbo codes made of 4-state RSC encoders of rate $1/2$ with generator matrix G_1, and the relay uses the same RSC encoder. For

OMARC/JNCC, we first imposed on the sources the use of the same signal sets. In this case, the two sources transmit their information symbols without any coding, while the relay uses the 4-state RSC encoder of rate $1/2$. The corresponding results demonstrated considerable gains in favor of our approach. We next carried out another experience, where constellation expansion is employed for OMARC, as explained in the outage comparisons. Thus, in the case of OMARC with $\eta = 4/3$ b./c.u., both sources use the same turbo code as the SOMARC with 16QAM modulation, and the relay uses the 4-state RSC encoder of rate $1/2$ with QPSK constellation. Similarly, in the case of $\eta = 8/3$ b./c.u., both sources use the turbo code made of 4-state RSC encoders of rate $1/2$ with generator matrix $\mathbf{G}_1$, whose parity bits are punctured to result in a code of rate $2/3$. They then use 64QAM constellation. The relay uses the same RSC encoder as the previous case with 16QAM modulation. The corresponding results are depicted in Figure 7.15 for the spectral efficiency of $\eta = 4/3$ b./c.u., and in Figure 7.16 for the spectral efficiency of $\eta = 8/3$ b./c.u., for both $N_D = 1$ and $N_D = 4$. Here again, the SOMARC outperforms the OMARC in most cases and the performance gains are considerable for $N_D = 4$. The exception is the case of $\eta = 8/3$ b./c.u. and for $N_D = 1$, where the SOMARC starts to outperform the OMARC with constellation expansion at a relatively high SNR ($\gamma = 24$ dB).

To pursue our comparison of practical designs, we compare the individual BLER of SOMARC/JNCC with the individual BLER of SOMARC/SNCC. In SOMARC/SNCC, both sources use the same punctured turbo code as the SOMARC/JNCC, and the relay, as previously mentioned, has two choices: (1) it uses the same input alphabet as the case of SOMARC/JNCC and transmits its information symbols without any coding; (2) it performs constellation expansion. In case (2), for $\eta = 4/3$ b./c.u., the relay uses the same punctured turbo code as the sources with 16QAM modulation, and for $\eta = 8/3$ b./c.u., it uses the punctured turbo code of rate 2/3 made of 4-state RSC encoders of rate 1/2 with generator matrix $\mathbf{G}_1$. It then uses 64QAM constellation. The corresponding results are depicted in Figure 7.17 for the spectral efficiency of $\eta = 4/3$ b./c.u., and in Figure 7.18 for the spectral efficiency of $\eta = 8/3$ b./c.u., for both $N_D = 1$ and $N_D = 4$. As we see, the SOMARC/JNCC outperforms the SOMARC/SNCC, and the power gains are approximately the same as the ones predicted by theoretical bounds.

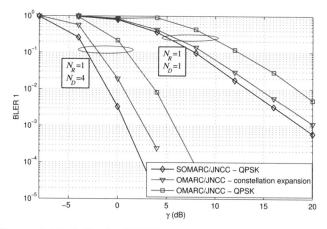

Figure 7.15. *Individual BLER (e.g. for S_1) – SOMARC/JNCC versus OMARC/JNCC – $\eta = 4/3$ b./c.u.*

7.7. Conclusion

We have introduced a new class of MARC, referred to as SOMARC, and we have analyzed it from both an information-theoretic and a practical code design perspective. We have derived the SOMARC individual information outage probability, conditional on JNCC and SNCC. We have also presented new JNCC schemes flexible in terms of number of sources, encoders and modulations. For the 2-source symmetric case and targeted sum rates $\eta = 4/3$ b./c.u. and $\eta = 8/3$ b./c.u., we have shown that our proposed schemes are more efficient than (1) conventional distributed JNCC for OMARC and (2) conventional SNCC schemes. Moreover, the proposed SOMARC/JNCC performs very close to the outage limit for both cases of single and multiple receive antennas at the destination, and for the fixed sum rate of $\eta = 4/3$ b./c.u. We have verified that the semi-orthogonal multiple access exhibits considerable gains over orthogonal multiple access, even in the case of a single receive antenna at the destination. There are several open doors for future works. For example, practical design and theoretical bounds need to be investigated for the more complex non-orthogonal MARC (NOMARC) where the constraint (3) that the sources remain silent during the transmission time slot of the relay is removed. Similarly, in a second step, the half-duplex constraint of the relay itself could be also relaxed. Finally, better relaying functions than SDF could be investigated, especially in the context of SOMARC.

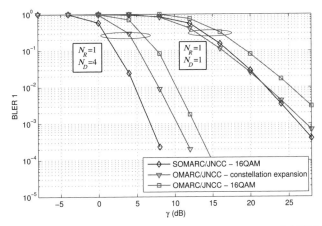

Figure 7.16. *Individual BLER (e.g. for S_1) – SOMARC / JNCC versus OMARC / JNCC – $\eta = 8/3$ b. / c.u.*

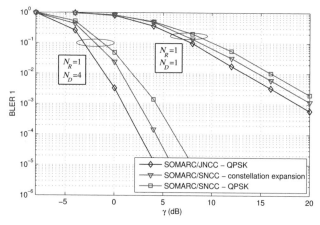

Figure 7.17. *Individual BLER (e.g. for S_1) – SOMARC/JNCC versus SOMARC/SNCC – $\eta = 4/3$ b./c.u.*

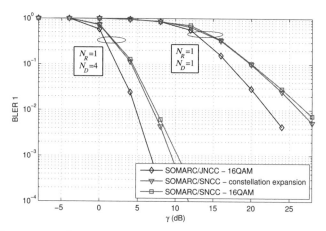

Figure 7.18. *Individual BLER (e.g. for S_1) – SOMARC/JNCC versus SOMARC/SNCC – $\eta = 8/3$ b./c.u.*

7.8. Appendix. MAC outage performance at high SNR

We want to prove that, in the large SNR regime, and for the special case of one receive antenna, the outage probability of an M-user slow fading MAC (with Gaussian signaling) behaves as that of an orthogonal MAC. Here, we refer to as outage probability of an M users' MAC, the probability that at least one user out of M is in outage. The M users are transmitting at the same rate R b./c.u., and the received instantaneous SNR of user i, $i \in \{1, \ldots, M\}$, is $|h_i|^2 P/N_0$, where h_i are independent channel fading coefficients following the pdf $\mathcal{CN}(0, 1)$. The outage probability of the considered MAC is defined as:

$$p_{out}^{MAC} = \Pr \left\{ \log \left(1 + \frac{P \sum_{i \in S} |h_i|^2}{N_0} \right) < |S|R, \quad \forall S \subseteq \{1, \ldots, M\} \right\}$$

[7.41]

where $|S|$ is the cardinality of the set S. Under independent Rayleigh fading, $h = \sum_{i \in S} |h_i|^2$ is the sum of $|S|$ i.i.d exponential random variables with parameter 1 and is distributed as

$$f(h) = \frac{1}{(|S| - 1)!} h^{|S|-1} e^{-h} , \quad h \geq 0.$$

[7.42]

For a fixed R and very high $\gamma = P/N_0$, $\Pr\{h < 2^{|S|R} - 1/\gamma\}$ is always lower than that of the case $|S| = 1$. Thus, at very high γ, the dominating events for p_{out}^{MAC} are for $|S| = 1$, which corresponds to an orthogonal MAC outage event. Thus

$$p_{out}^{MAC} \simeq \Pr \left\{ \log \left(1 + \frac{P|h_i|^2}{N_0} \right) < R, \quad \forall i \in \{1, \cdots, M\} \right\}$$

$$= M \Pr \left\{ |h_1|^2 < \frac{2^R - 1}{\gamma} \right\},$$

[7.43]

where the equality in [7.43] follows from the fact that the exponential random variables $|h_i|^2$, $i \in \{1, \cdots, M\}$, are independent and have the same Cumulative Distribution Function (CDF). Since, $h = |h_1|^2$ is distributed as $f(h) = e^{-h}$ for $h \geq 0$, it yields

$$p_{out}^{MAC} \simeq M \left(1 - \exp \left(\frac{-(2^R - 1)}{\gamma} \right) \right)$$

[7.44]

which can be approximated at high SNR as:

$$p_{out}^{\text{MAC}} \simeq M\frac{2^R - 1}{\gamma}. \tag{7.45}$$

It confirms that the MAC outage probability at high SNR decays as $1/\gamma$ similarly to the case of orthogonal MAC or single-user interference free channels. As a result, the probability of having n users ($1 \leq n \leq M$) in outage decays as γ^{-n}.

7.9. Bibliography

[AHL 00] AHLSWEDE R., CAI N., LI S.-Y.R., YEUNG R.W., "Network information flow", *IEEE Transaction on Information Theory*, vol. 46, pp. 1204–1216, July 2000.

[BAH 74] BAHL L., COCKE J., JELINEK F., RAVIV R., "Optimal decoding of linear codes for minimizing symbol error rate", *IEEE Transaction on Information Theory* vol. 20, pp. 284–286, March 1974.

[BAO 08] BAO X., LI J., "Adaptive network coded cooperation (ANCC) for wireless relay networks: matching code-on-graph with network-on-graph", *IEEE Transaction on Wireless Communications* vol. 7, no. 2, pp. 574–583, February 2008.

[CAI 98] CAIRE G., TARICCO G., BIGLIERI E., "Bit-interleaved coded modulation", *IEEE Transaction on Information Theory*, vol. 44, no. 3, pp. 927–946, May 1998.

[CHE 06] CHEN Y., KISHORE S., LI J., "Wireless diversity through network coding", *Proceedings of IEEE WCNC'06*, vol. 3, Monticello, USA, pp. 1681–1686, April 2006.

[DUY 10] DUYCK D., CAPIRONE D., MOENECLAEY M., BOUTROS J., "Analysis and construction of full-diversity joint network-LDPC codes for cooperative communications", *EURASIP Journal on Wireless Communications and Networking*, vol. 2010, pp. 1–16, January 2010.

[GRA 89] GRAHAM R.L., KNUTH D.E., PATASHNIK O., *Concrete Mathematics*, Addison-Wesley, Reading, USA, 1989.

[GUO 07] GUO Z., WANG B., CUI J.-H., "Efficient error recovery using network coding in underwater sensor networks", *Proceedings of 6th International IFIP-TC6 Conference on Networking '07*, Atlanta, USA, pp. 227–238, May 2007.

[HAG 96] HAGENAUER J., OFFER E., PAPKE L., "Iterative decoding of binary block and convolutional codes", *IEEE Transaction on Information Theory*, vol. 42, no. 2, pp. 429–445, September 1996.

[HAG 97] HAGENAUER J., "The turbo principle: tutorial introduction and state of the art", *Proceedings of the 1st International Symposium on Turbo Codes*, Brest, France, pp. 1–12, September 1997.

[HAT 10a] HATEFI A., VISOZ R., BERTHET A., "Joint channel-network coding for the semi-orthogonal multiple access relay channel", *Proceedings of IEEE VTC-Fall '10*, Ottawa, Canada, September 2010.

[HAT 10b] HATEFI A., VISOZ R., BERTHET A., "Joint channel-network turbo coding for the non-orthogonal multiple access relay channel", *Proceedings of IEEE PIMRC '10*, Istanbul, Turkey, September 2010.

[HAT 11] HATEFI A., VISOZ R., BERTHET A., "Full diversity distributed coding for the multiple access half-duplex relay channel", *Proceedings of IEEE Netcod '11*, Beijing, China, July 2011.

[HAU 05] HAUSL C., SCHRECKENBACH F., OIKONOMIDIS I., BAUCH G., "Iterative network and channel coding on a Tanner graph", *Proceedings of Annual Allerton Conference on Communication, Control and Computing*, Monticello, USA, September 2005.

[HAU 06] HAUSL C., DUPRAZ P., "Joint network-channel coding for the multiple access relay channel", *Proceedings of the 3rd Annual IEEE Communications Society on Sensor and Ad Hoc Communications and Networks*, vol. 3, pp. 817–822, September 2006.

[KAT 05] KATTI S., KATABI D., HU W., RAHUL H., MEDARD M., "The importance of being opportunistic: practical network coding for wireless environments", *Proceedings of the 43rd Allerton Conference '05*, September 2005.

[KAT 06] KATTI S., RAHUL H., HU W., KATABI D., MEDARD M., CROWCROFT J., "XORs in the air: practical wireless network coding", *Proceedings of SIGCOMM '06*, Pisa, Italy, September 2006.

[KRA 00] KRAMER G., VAN WIJNGAARDEN A.J., "On the white Gaussian multiple access relay channel", *Proceedings of IEEE ISIT'00*, Sorrento, Italy, June 2000.

[MAL 99] MALKAMÄKI E., LEIB H., "Coded diversity on block-fading channel", *IEEE Transaction on Information Theory*, vol. 45, no. 2, pp. 771–781, March 1999.

[SAN 04] SANKARANARAYANAN L., KRAMER G., MANDAYAM N., "Capacity theorems for the multiple-access relay channel", *Proceedings of Annual Allerton Conference on Communication, Control and Computing*, Monticello, USA, September 2004.

[RAP 98] RAPHAELI D. and ZARAI Y., "Combined turbo equalization and turbo decoding", *IEEE Communications Letters*, vol. 2, pp. 107–109, April 1998.

[TSE 05] TSE D., VISWANATH P., *Fundamentals of Wireless Communication*, Cambridge University Press, Cambridge, 2005.

[UNG 82] UNGERBOECK G., "Channel coding with multilevel/phase signals", *IEEE Transaction on Information Theory*, vol. 28, pp. 55–67, July 1982.

[WAN 08] WANG T., GIANNAKIS G., "Complex field network coding for multiuser cooperative communications", *IEEE Journal on Selected Areas in Communications*, vol. 26, no. 3, pp. 561–571, April 2008.

[WOL 07] WOLDEGEBREAL D.H., KARL H., "Multiple-access relay channel with network coding and non-ideal source-relay channels", *Proceedings of the 4th International Symposium on Wireless Communication Systems*, Trondheim, Norway, October 2007.

[WU 05] WU Y., CHOU P.A., KUNG S. Y., "Information exchange in wireless networks with network coding and physical-layer broadcast", *Proceedings of the 39th Annual Conference on Information Sciences and Systems (CISS)*, Baltimore, USA, March 2005.

[YAN 07] YANG S., KOETTER R., "Network coding over a noisy relay: a belief propagation approach", *Proceedings of IEEE ISIT'07*, Nice, France, June 2007.

[ZEI 08] ZEITLER G., KOETTER R., BAUCH G., WIDMER J., "Design of network coding functions in multihop relay networks", *Proceedings of the 5th International Symposium on Turbo Codes and Related Topics*, Lausanne, Switzerland, September 2008.

[ZEI 09] ZEITLER G., KOETTER R., BAUCH G., WIDMER J., "On quantizer design for soft values in the multiple access relay channel", *Proceedings of IEEE ICC'09*, Lausanne, Switzerland, September 2009.

Chapter 8

Robust Network Coding

Network coding [AHL 00, LI 03], and more specifically random linear network coding [KOE 03, CHO 03, HO 03, HO 06], is a powerful tool for delivering information across a network. Random coding techniques may be implemented in a distributed way within network elements, independently of the structure of the network. In [HO 06], it has been shown that the max-flow capacity of the network can be reached with probability exponentially approaching one, with the size of the Galois field in which the random coding operations are performed. This work has led to a number of practical schemes such as COPE, ANC, MIXIT, and MORE [KAT 05, KAT 07b, KAT 07a].

Nevertheless, network coding is very sensitive to transmission errors, packet losses, and corrupted packets, which are intentionally injected by malicious nodes. Recombinations carried out by each node lead to a progressive contamination of the set of clean packets by the erroneous

Chapter written by Lana IWAZA, Marco Di RENZO and Michel KIEFFER.

ones, which makes the decoding impossible at the receiver side. On the other hand, even in the absence of errors, losses of packets lead to an insufficient number of packets at the receiver side, making the use of already received packets impossible.

Error-correcting network coding techniques aim at protecting packets from transmission errors, from erroneous packets, and/or losses. Error-correcting network coding techniques introduce a certain level of *redundancy* and are similar in principle to classical error-correcting codes. We can distinguish between the two families of codes. The codes introduced in [CAI 02, ZHA 08] focus both on network coding and on the introduction of redundancy. These codes require an *a priori* knowledge of the architecture of the network and the way in which network coding is carried out, see section 8.1 for further details. These results are extended to the framework of random network coding in [HO 06, BAL 09], see section 8.2. The techniques introduced in [JAG 07, KOE 08, SIL 08, AHL 09] exploit the fact that in the absence of errors, random network coding preserves the vector space spanned by the transmitted packets. The proposed robust network codes have properties that are relatively independent from the way the network coding is carried out, see section 8.3.

Joint decoding techniques exploit the existing redundancy in the communication networks [DUH 09]. In the case of *joint channel–network decoding* [HAU 06b, THO 08], temporal or spatial diversity or the presence of channel codes [GUO 09, KLI 07] is used to combat the noise introduced by the communication channels, in particular wireless channels; see section 8.4. *Joint source–network decoding* allows the recovery of all or part of the initial packets, even in the presence of an insufficient number of received packets, by exploiting the correlation between transmitted data packets; see section 8.5.

Therefore, these techniques provide a certain robustness against packet loss.

8.1. Coherent network error-correction codes

The notations and the content of this section are largely inspired by [ZHA 11, YAN 11, ZHA 08]. For this type of network error-correcting codes, the topology of the network and the considered network code are assumed to be known by each destination node [YEU 06, CAI 06].

A communication network is described by a directed acyclical graph $\mathcal{G} = \{\mathcal{V}, \mathcal{E}\}$. A link $e = (i, j) \in \mathcal{E}$ represents a channel linking the nodes $i \in \mathcal{V}$ and $j \in \mathcal{V}$. The set of links emerging from a node $i \in \mathcal{V}$ is written as $O(i)$, and the set of paths converging at i is written as $I(i)$. A multicast network is a triple $(\mathcal{G}, s, \mathcal{T})$, where $\mathcal{G}$ is a network, $s \in \mathcal{V}$ is the source, and $\mathcal{T}$ is the set of destination nodes. We assume that $I(s) = \emptyset$, $O(t) = \emptyset$ for every $t \in \mathcal{T}$. Let $n_s = |O(s)|$. Subsequently, $\mathbb{F}$ represents the Galois field with q elements.

The source node s encodes the message to transmit as a row vector $\mathbf{x} = [x_1, \ldots, x_{n_x}] \in \mathbb{F}^{n_x}$ called a *codeword*. The set of codewords is written as $\mathcal{C}$. Each component of $\mathbf{x}$ is therefore sent on one of the links of $O(s)$. An error vector $\mathbf{z} \in \mathbb{F}^{|\mathcal{E}|}$ allows us to describe the errors introduced by the links in the network. If we denote $\bar{f}_e$ and f_e as the input and output of the link e and if an error z_e is introduced on the link $e \in \mathcal{E}$, then $f_e = \bar{f}_e + z_e$. For every subset of links $\rho \in \mathcal{E}$, we introduce the two vectors $\mathbf{f}_\rho = [f_e, e \in \rho]$ and $\bar{\mathbf{f}}_\rho = [\bar{f}_e, e \in \rho]$. A code for the network $\mathcal{G}$ is therefore defined by a set of codewords $\mathcal{C} \subset \mathbb{F}^{n_s}$ and a family of local coding functions $\{\bar{\beta}_e, e \in \mathcal{E} \backslash O(s)\}$, with $\bar{\beta}_e : \mathbb{F}^{|I(\text{source}(e))|} \to \mathbb{F}$ such that

$$\bar{f}_e = \bar{\beta}_e(F_{I(\text{source}(e))}) \qquad\qquad [8.1]$$

and where source(e) indicates the node from which e emerges. Assume that the destination node t receives the vector $\mathbf{u}_t = (u_e,\ e \in I(t))$. An iterative application of [8.2] allows us to express $\mathbf{u}_t$ as a function of $\mathbf{x}$ and of the error vector $\mathbf{z}$

$$\mathbf{u}_t = F_{st}(\mathbf{x}, \mathbf{z}) \qquad\qquad [8.2]$$

where $F_{st}(\mathbf{x}, \mathbf{z})$ represents the set of network coding operations taking place between the source s and destination t. In the case of coherent network codes, the structure of $F_{st}(\mathbf{x}, \mathbf{z})$ is assumed to be known at the decoder and is used to perform the estimation of $\mathbf{x}$ from $\mathbf{u}_t$. To characterize the error-correction capacity of a network code, it is necessary to introduce the notion of distance between codewords [ZHA 08]. For this purpose, consider the set of vectors that can be received by the node t when the source transmits a codeword $\mathbf{x}$ and the network introduces an error vector $\mathbf{z}$ with a Hamming weight $w_H(\mathbf{z})$ less than c

$$\Phi_t(\mathbf{x}, c) = \{F(\mathbf{x}, \mathbf{z}) \text{ st } w_H(\mathbf{z}) \leqslant c\} \qquad\qquad [8.3]$$

It is possible to deduce from $\Phi_t(\mathbf{x}, c)$ a pseudo-distance between two codewords $\mathbf{x}$ and $\mathbf{y}$ emitted by the source

$$D_t(\mathbf{x}, \mathbf{y}) = \min\{c_1 + c_2 \text{ st } |c_1 - c_2| \leqslant 1 \text{ and }$$
$$\Phi_t(\mathbf{x}, c_1) \cap \Phi(\mathbf{x}, c_2) \neq \emptyset\} \qquad\qquad [8.4]$$

and a minimal distance for the network code at the node t

$$d_{\min,t} = \min\{D_t(\mathbf{x}, \mathbf{y}), \mathbf{x} \neq \mathbf{y}\} \qquad\qquad [8.5]$$

The decoder seeking the minimum weight error vector (maximum likelihood decoder if all codewords have same probability) can therefore be constructed in the following way. First, we search for the set $\mathcal{P}$ of pairs $(\mathbf{x}, \mathbf{z})$ satisfying [8.2]. In the subset $\mathcal{P}_w \subset \mathcal{P}$ of the pairs $(\mathbf{x}, \mathbf{z})$ whose Hamming weight $\mathbf{z}$ is minimal, if all the pairs have same $\mathbf{x}$, then the error is said to be *correctable* and x is the estimation of the transmitted message. If this is not the case, the error is not *correctable*. It has been shown in [YAN 08] that the correction capacity of a network code (with a decoder that searches for the minimum weight error vector) is $\lfloor (d_{\min} - 1)/2 \rfloor$, where $\lfloor \cdot \rfloor$ indicates the rounding toward $-\infty$. In the case of linear network codes, the functions $\bar{\beta}_e$ are linear and for every $e \in \mathcal{E} \backslash O(s)$, we have

$$\bar{f}_e = \sum_{e' \in \mathcal{E}} \beta_{e',e} F_{e'} \tag{8.6}$$

where $\beta_{e',e}$ is the local coding coefficient of the node e' toward the node e. Using [8.6], [KOE 03] has shown that [8.2] can be written as

$$\mathbf{u}_t = \mathbf{x} \mathbf{F}_{s,t} + \mathbf{z} \mathbf{F}_t \tag{8.7}$$

where $\mathbf{F}_{s,t}$ and $\mathbf{F}_t$ can be deduced from [8.2] and are perfectly known. In the case of linear network codes, [8.4] becomes

$$D_t(\mathbf{x}, \mathbf{y}) = \min \{c \text{ st } (\mathbf{x} - \mathbf{y}) \mathbf{F}_{s,t} \in \Phi_t(c)\} \tag{8.8}$$

with

$$\Phi_t(c) = \{\mathbf{z} \mathbf{F}_t, \ \mathbf{z} \in \mathbb{F}^{|\mathcal{E}|}, \ w_H(\mathbf{z}) \leqslant c\} \tag{8.9}$$

the set of messages received when the zero codeword is sent. The main bounds in terms of error-correction codes have been extended to network codes in [YEU 06, CAI 06, YAN 11] such as the Hamming, the Singleton, and the Gilbert–Varsamov bounds, as well as in [BYR 08] for the Plotkin and Elias bounds. For the Hamming and Singleton bounds

$$d_{\min} = \min_{t \in \mathcal{T}} d_{\min,t} \tag{8.10}$$

and

$$n = \min_{t \in \mathcal{T}} \mathbf{maxflow}(s, t) \tag{8.11}$$

In the case of a network code for which $\text{rank}(\mathbf{F}_{s,t}) = r_t$ and $d_{\min,t} > 0$, the Hamming bound may be written as

$$|\mathcal{C}| \leqslant \min_{t \in \mathcal{T}} \frac{q^{r_t}}{\sum_{i=0}^{\tau_t} \binom{r_t}{i} (q-1)^i} \qquad [8.12]$$

with $\tau_t = \lfloor (d_{\min,t} - 1)/2 \rfloor$. The Singleton bound becomes

$$|\mathcal{C}| \leqslant q^{r_t - d_{\min,t} + 1} \qquad [8.13]$$

for every node t, see [YAN 11]. The Singleton bound [8.13] allows us to extend the notion of *maximum distance separable* (MDS) codes to network codes [YEU 05]. A network code where the Singleton bound is reached is said to be MDS. It is optimal in the sense that it exploits all the redundancy in the network error-correcting code. A code construction method enabling the Singleton bound [8.13] to be reached has been proposed in [YAN 11]. The technique consists of first constructing the local coding coefficients, which ensure that the rank of matrices $\mathbf{F}_{s,t}$ is always sufficient. This can be done using the Jaggi–Sanders algorithm [JAG 05]. The codewords are then generated so that there is sufficient distance between them regardless of which destination node t is considered. The associated decoding algorithms are presented in [YEU 06, CAI 06]. These techniques will be described in further detail in the rest of this chapter. See also [MAT 07, ZHA 08] as well as [BAL 09] for further details on this type of codes.

8.2. Codes for non-coherent networks, random codes

The random network codes proposed in [HO 03, HO 06] can be seen as a practical solution to network coding, which can easily adapt to variations in the network topology since they are decentralized. In the case of random coding, the matrices $\mathbf{F}_{s,t}$ and $\mathbf{F}_t$ introduced in [8.7] are random. While it is possible to deduce $\mathbf{F}_{s,t}$ from the received packet headers (assuming that they have not been corrupted), $\mathbf{F}_t$, on the other hand, cannot be easily deduced. In the absence of transmission errors, the probability that a destination node $t \in \mathcal{T}$ is not capable of decoding the message received can be expressed as a function of the rank of $\mathbf{F}_{s,t}$

$$P_e^{(t)} = \Pr(\mathrm{rank}(\mathbf{F}_{s,t}) < n_x) \tag{8.14}$$

The probability that at least one of the destination nodes is incapable of decoding the received message is deduced from [8.14]

$$P_e = \Pr(\exists t \in \mathcal{T} \text{ st } \mathrm{rank}(\mathbf{F}_{s,t}) < n_x) \tag{8.15}$$

see [HO 06]. If c_t denotes the min-cut capacity between s and t, then $\delta_t = c_t - n_x$ corresponds to the redundancy at t. The probability of errors at the receiver t is therefore bounded as follows

$$P_e^{(t)} \leqslant 1 - \sum_{i=n_x}^{n_x+\delta_t} \binom{n_x + \delta_t}{i} \left(1 - p - \frac{1-p}{q}\right)^{Li}$$
$$\times \left(1 - \left(1 - p - \frac{1-p}{q}\right)^L\right)^{n_x+\delta_t-i} \tag{8.16}$$

where L indicates the length of the longest path between s and t and p is the link erasure probability. When the links are perfectly reliable ($p = 0$), [8.16] becomes

$$P_e^{(t)} \leqslant 1 - \sum_{i=0}^{\delta_t} \binom{C_t}{i} \left(1 - \frac{1}{q}\right)^{L(C_t-i)} \left(1 - \left(1 - \frac{1}{q}\right)^L\right)^i \tag{8.17}$$

In the presence of errors, the results of [ZHA 08] briefly presented in section 8.1 can be extended. However, for a given code, the minimum distance $d_{\min,t}$ introduced in [8.5] becomes a random variable $D_{\min,t}$. Once the code $\mathcal{C}$ is fixed, the distance

$d_{\min,t}$ will depend on the (random) elements of $\mathbf{F}_{s,t}$. A partial characterization of $D_{\min,t}$ has been proposed in [BAL 09]

$$\Pr(D_{\min,t} < \delta_t + 1 - d) \leqslant \frac{\binom{|\mathcal{E}|}{\delta_t - d}\binom{d+|\mathcal{J}|+1}{|\mathcal{J}|}}{(q-1)^{d+1}} \qquad [8.18]$$

where $\mathcal{J} \subset \mathcal{E}$ is the set of internal nodes in the network. This result allows us to deduce the probability of existence of an MDS code according to the size q of the Galois field in which the coding operations take place, see [BAL 09] for further details.

8.3. Codes for non-coherent networks, subspace codes

The network coding error-correcting techniques proposed in [KOE 08, SIL 08] are very different from the ones previously introduced. A non-coherent network model is considered, where neither the coder nor the decoder need to know the topology of the network nor the way in which combinations of packets are carried out. This work is motivated by the fact that, in the absence of errors, network coding preserves the vector space spanned by the transmitted packets. The coding operation is carried out via the transmission of a vector space inside a set of possible vector spaces (which represents the set of codewords). A destination node must identify the vector subspace belonging to the code found to be the closest (in a sense to be defined) to the vector space spanned by the received packets. The received vector space can be different from the one that has been transmitted, depending on the packet losses, transmission errors, or erroneous packets deliberately injected by malicious nodes.

8.3.1. *Principle of subspace codes*

In this approach, the transmission of information from the source s to a destination node t is conveyed by the injection into the network of a vector subspace $V \subset \mathbb{F}^n$ and by the reception of a subspace $U \subset \mathbb{F}^n$. Let $\mathbf{x} = \{\mathbf{x}_1, \ldots, \mathbf{x}_{n_s}\}$, with $\mathbf{x}_i \in \mathbb{F}^n$, be the set of vectors (data packets) injected by the source s and forming a base of V. In the absence of errors, $t \in T$ receives a set of packets $\mathbf{u} = \{\mathbf{u}_1, \ldots, \mathbf{u}_{n_t}\}$ formed by linear combinations of $\{\mathbf{x}_1, \ldots, \mathbf{x}_{n_s}\}$, such that $\mathbf{u}_j = \sum_{i=1}^{n_s} h_{ji} \mathbf{x}_i$, where the h_{ji} are random coefficients of $\mathbb{F}$. The effect of potential transmission errors is modeled by the introduction of *packets* of errors $\mathbf{z} = \{\mathbf{z}_1, \ldots, \mathbf{z}_{n_z}\}$ throughout the network. Since these packets can be injected into any link or node in the network, at receiver side, we get

$$\mathbf{u}_j = \sum_{i=1}^{n_s} h_{ji} \mathbf{x}_i + \sum_{k=1}^{n_z} g_{jk} \mathbf{z}_k \qquad [8.19]$$

where the $g_{jk} \in \mathbb{F}$ are again random. In matrix form, we obtain

$$\mathbf{u} = H\mathbf{x} + G\mathbf{z} \qquad [8.20]$$

Model [8.19] is close to [8.7], but in [8.7], symbols belonging to $\mathbb{F}$ are transmitted while in [8.19] packets are sent through the network. In [8.7], $\mathbf{F}_{s,t}$ and $\mathbf{F}_t$ are perfectly known when the network structure and the network coding operations are known, which is not the case with the coefficients h_{ji} and g_{jk} (this is why we consider here non-coherent network codes). With this type of model, the aim of the receiver cannot be to precisely identify $\mathbf{x}$, but rather to identify the vector subspace V spanned by the vectors of $\mathbf{x}$, based on the knowledge of the vector subspace U created by the elements of $\mathbf{u}$. To introduce the notion of subspace codes, we consider a vector space W of dimension n on $\mathbb{F}$, for example $\mathbb{F}^n$. $\mathcal{P}(W)$ is the set of all the vector subspaces of W. The dimension of a subspace $V \in \mathcal{P}(W)$ is written as $\dim(V)$. We can show that [KOE 08] for every $A \in \mathcal{P}(W)$ and $B \in \mathcal{P}(W)$,

$$d(A, B) = \dim(A + B) - \dim(A \cap B) \qquad [8.21]$$

is a distance between vector subspaces. A subspace code is therefore a subset of $\mathcal{C} \subset \mathcal{P}(W)$. A *codeword* of $\mathcal{C}$ is a vector subspace of $\mathcal{C}$. The minimum distance of $\mathcal{C}$ is the minimum distance between two distinct codewords while using the distance [8.21]

$$d_{\min}(\mathcal{C}) = \min_{X,Y \in \mathcal{C}, \; X \neq Y} d(X,Y) \qquad [8.22]$$

The maximum dimension of the codewords of $\mathcal{C}$ is $\ell(\mathcal{C}) = \max_{X \in \mathcal{C}} \dim(X)$. When the dimension of all the codewords of $\mathcal{C}$ is the same, the code is of constant dimension. Assume that a codeword $V \in \mathcal{C}$ is sent by the source and that U is received by a destination $t \in \mathcal{T}$, it is possible to describe the behavior of the network as

$$U = H_k(V) \oplus Z \qquad [8.23]$$

with $k = \dim(U \cap V)$, $H_k(V)$ is a subspace of V with dimension k such that $H_k(V) \cap V = 0$. This type of model illustrates the impact of network coding and the introduction of errors in terms of operations on vector subspaces. With this model, the network introduces $\rho = \dim(V) - k$ cancellations and $n_z = \dim(Z)$ errors. In this case, [KOE 08] shows that if $2(n_z + \rho) < d_{\min}(\mathcal{C})$, then a decoder with a minimum distance allows getting V from U. A generalization of the Singleton bound is proposed for these codes [KOE 08]. A construction of codes on subspaces similar to Reed–Solomon codes allowing the Singleton bound to be reached as well as a decoding algorithm with minimum distance for this family of codes is detailed in [KOE 08], emphasizing constant dimension codes.

8.3.2. *Recent developments*

This research has led to a number of recent developments. Constant dimension codes are studied in [XIA 08] and applied to network coding by demonstrating that Steiner structures are optimal constant dimension codes. Johnson-type bounds are also calculated. In [GAB 08], several new codes and bounds exploiting the distance between subspaces [8.21] are explored. In [SIL 08], a wide class of constant dimension codes is studied and a new distance considering the rank metrics is introduced. Codes associated with this metric are introduced and an effective decoding algorithm for this family is proposed. Several constant dimension codes are introduced in [KOH 08], with a larger number of codewords than in the case of the previously examined codes. Performance bounds as well as construction methods for the code family introduced in [KOE 08] are proposed in [AHL 09]. An analysis of the geometric properties of the codes using rank metric is carried out in [GAD 09]. The lower and upper bounds of the cardinality of codes of a given rank are evaluated, which enables an analysis of the performance of these codes. In [ETZ 09], a new multilevel approach examining the construction of subspace codes is presented. The authors show that the codes proposed in [KOE 08] represent a specific case of the proposed family of codes. A Gilbert–Varshamov bound relative to the codes constructed in [SIL 09] is introduced in [KHA 09], exploiting the injection metric. Finally, [CHE 09] studies the practical implantation of the codes introduced in [KOE 08]. The construction of these codes for small Galois fields and limited error-correction capacity is feasible, and improves the network performance in terms of throughput.

8.4. Joint network–channel coding/decoding

This section aims to show how, in a wireless context, the redundancy existing at the network level can help improve channel decoding performance by performing joint network–channel decoding. This joint approach allows reducing the number of packets lost due to transmission errors on wireless networks. This is achieved by using, on the one hand, the network spatial diversity and, on the other hand, the redundancy introduced by channel codes on the low layers of communication protocols. This research is motivated by [EFF 03], which highlights the limits of the coding approaches in which the network and the channel or the source and the network are separated. Studies carried out on canonical networks demonstrate that source–network separation remains valid for some networks although this is not the case for network–channel separation. In [LEE 07], it is also shown that despite the fact that separation remains valid in some cases, a separate processing, for example source–network, results in higher costs, for example in terms of bandwidth or energy, than in a joint treatment.

Wireless networks are a favored area of application for joint network–channel decoding techniques. In contrast to wired networks where lower layers of the protocol stack are supposed to provide error-free links, wireless networks provide packets that may be erroneous. Joint network–channel decoding techniques exploit the redundancy introduced by the network coding operations to improve the capacity of the channel code to correct the transmission errors. Instead of focusing on guaranteeing an error-free transmission on each link, we are more interested in guaranteeing error-free decoding at the destination nodes. The latter use the data received from incoming links for decoding. In the presence of links providing a certain level of redundancy, error-free decoding is possible even if decoding at the level of each individual link is not possible. Joint network–channel decoding is therefore only useful when network coding introduces redundancy. The first practical application of this concept to networks with relays has been proposed in [HAU 06b]. Iterative network–channel decoding methods for relay networks as well as for multiple access relay channels have been proposed in [HAU 06a] and [HAU 06b].

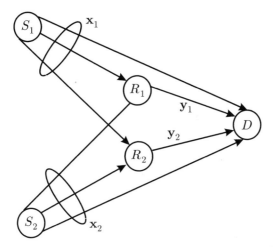

Figure 8.1. *Wireless networks with two sources, two relays,
and one destination*

8.4.1. *Principle*

Consider a wireless network topology consisting of two
sources S_1 and S_2, two intermediate relay nodes R_1 and R_2,
and a destination node D, see Figure 8.1. The sources generate
two information messages x_1 and x_2 of k symbols each, and
protect them using channel codes to obtain two independent
packets with n symbols each, p_1 and p_2, which are then
transmitted toward D. The relays receive the two packets,
process them, and retransmit them to D. To simplify, the links
are assumed to be without errors and the communications
are carried out on two orthogonal channels where mutual
interference is negligible. As a result, D receives four packets
from which it attempts to recover the information messages
x_1 and x_2 sent by the sources. Assume that the two packets p_1
and p_2 can be expressed as a function of x_1 and x_2 as follows:

$$p_1 = x_1 G_1 \text{ and } p_2 = x_2 G_2 \qquad [8.24]$$

where G_1 and G_2 are two channel coding matrices with
dimension $k \times n$ and elements belonging to $\mathbb{F}$, a Galois field

with q elements. The relay nodes R_1 and R_2 directly receive $\mathbf{p}_1$ and $\mathbf{p}_2$ and are therefore capable of decoding them to obtain $\mathbf{x}_1$ and $\mathbf{x}_2$, which are then re-encoded using a channel code and a network code to obtain

$$\mathbf{y}_1 = a_{11}\mathbf{x}_1 G_{11} + a_{12}\mathbf{x}_2 G_{12} \qquad [8.25]$$

and

$$\mathbf{y}_2 = a_{21}\mathbf{x}_1 G_{21} + a_{22}\mathbf{x}_2 G_{22} \qquad [8.26]$$

where the a_{ij} are network coding coefficients and where the matrices G_{ij} are channel code generator matrices used at the relay. As a result, D receives four packets $\mathbf{p}_1$, $\mathbf{p}_2$, $\mathbf{y}_1$ and $\mathbf{y}_2$ from which it has to estimate $\mathbf{x}_1$ and $\mathbf{x}_2$ transmitted by the source. By adopting a matrix notation, we obtain the following equations

$$\begin{bmatrix} \mathbf{p}_1 \\ \mathbf{p}_2 \\ \mathbf{y}_1 \\ \mathbf{y}_1 \end{bmatrix} = \begin{bmatrix} G_1 & 0 \\ 0 & G_2 \\ a_{11}G_{11} & a_{12}G_{12} \\ a_{21}G_{21} & a_{22}G_{22} \end{bmatrix} \begin{bmatrix} \mathbf{x}_1 \\ \mathbf{x}_2 \end{bmatrix} = G_{\text{joint}} \begin{bmatrix} \mathbf{x}_1 \\ \mathbf{x}_2 \end{bmatrix}$$

$$[8.27]$$

where G_{joint} represents the generator matrix for the joint network–channel code. From [8.27], we see that channel and network codes can be considered as a unique code from the point of view of the network extremities and that the latter can be represented by a unique generator matrix G_{joint}. As a result, in the presence of transmission errors, messages $\mathbf{x}_1$ and $\mathbf{x}_2$ can be decoded at the destination by directly exploiting G_{joint} or by the use of an iterative decoding method, see for example [GUO 09].

8.4.2. *Recent developments*

Several studies linked to joint network–channel decoding have been proposed in [HAU 06a] and [HAU 06b]. These studies have focused on relaxing some of the hypotheses introduced in [HAU 06b], like assuming that the error correction is perfect between the source and the relays, see for example [BAO 06], [XIA 07], or [ALH 09]. Other results on joint network–channel decoding and more specifically code optimization can be found in [YIN 09].

8.5. Joint source–network coding/decoding

Joint source–network coding and decoding enable all or some of the packets transmitted by the sources to be recovered in the presence of an insufficient number of received network-coded packets, by exploiting the existing or artificially introduced correlation between the transmitted data packets. These techniques also enable the distributed compression of correlated messages generated by geographically distributed sources.

Regarding the robustness against losses, or capacity variations on some of the links of the network, an alternative solution to the network coding techniques presented in sections 8.1 to 8.3 consists of combining multiple description coding techniques [GOY 98], [GOY 01] and network coding. The aim is to exploit the redundancy introduced by these coding techniques to allow a progressive improvement of the quality of data reconstructed with the number of packets received at the receiver nodes [IWA 11]. An already existing correlation between data generated by the sources can also be exploited to obtain a scheme more robust to packet losses.

Regarding distributed compression, distributed source coding [SLE 73, WYN 76, CRI 05] can perform separate compression of correlated sources and may be as effective (when there are no losses) as joint compression. This technique is interesting in the case of sensor networks where it is possible to perform efficient compression even in the absence of coordination between the sensors [COL 09, HOW 08]. This solution does not, however, allow complete exploitation of the capacity of the network and assumes that the sensors have a precise estimate of the level of correlation between the data they produce. In this context, network coding is a natural solution for correlated data transmission on a network with diversity. The application of network coding for the compression of correlated sources has been proposed in [BAR 06, HO 04, WU 05] in the case of lossless coding. The proposed techniques provide effecient distributed algorithms, which are capable of exploiting diversity whether at the source or the channel level. In the case of coding with losses, *compressed sensing* [DON 06, CAN 06] allows an approximate reconstruction of the source by exploiting its properties of compressibility using random combinations of its samples. Network coding techniques inspired by compressed sensing have been proposed in [SHI 08], using network codes on the real fields. However, the data taken from wireless sensor networks are, in general, quantized and network coding in the case of real fields is therefore questionable.

8.5.1. *Exploiting redundancy to combat loss*

To combat packet losses in the network, it is possible to exploit the redundancy existing in the data generated by the source(s). This redundancy may be introduced artificially, as in [IWA 11], or be present naturally, as in [IWA 12].

8.5.1.1. *Artificially introduced correlation*

Two techniques for introducing redundancy are examined in [IWA 11]. The first involves a transformation matrix whose coefficients belong to $\mathbb{F}$ after quantization, see Figure 8.2.

The samples $\mathbf{x} \in \mathbb{R}^k$ generated by the source are quantized on q levels and are then transformed using a redundant transformation $T \in \mathbb{F}^{n \times k}$ with full rank k to obtain $\mathbf{z} = T\mathbf{y}$.

As a result, there exists a matrix $D \in \mathbb{F}^{(n-k) \times n}$ with full rank $n - k$ such that $Dz = 0$.

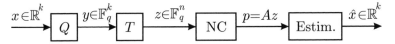

Figure 8.2. *Joint source–network coding with redundancy introduced by the transformation*

The elements of z are then transmitted in the network where the network coding operations, represented by the matrix A, are performed at the intermediate nodes. At the decoder side, the matrix

$$B = \begin{pmatrix} A \\ D \end{pmatrix}$$

is constructed from the received packets. If there exists a sub-matrix B' of B such that B' is of full rank k, then the elements of z can be reconstructed by a simple Gaussian elimination. This approach provides a good robustness against packet losses with a decoding complexity similar to that of classical network coding.

The second technique introduces redundancy via a frame expansion [GOY 01] of data generated by the source, see Figure 8.3.

Figure 8.3. *Joint source–network coding with redundancy introduced via a frame expansion*

The samples $x \in \mathbb{R}^k$ generated by the source are transformed using a frame expansion $F \in \mathbb{R}^{n \times k}$ to obtain $y = Fx \in \mathbb{R}^n$. A frame on $\mathbb{R}^k$ is a set of $n > k$ vectors $\{\varphi_i\}_{i=1...n}$

such that there is $B > 0$ and $C < \infty$ satisfying for every $\mathbf{x} \in \mathbb{R}^k$,

$$B\|x\|^2 \le \sum_{i=1}^{n} \langle x, \varphi_i \rangle^2 \le C\|x\|^2 \qquad [8.28]$$

where $\langle \cdot, \cdot \rangle$ is the scalar product of $\mathbb{R}^k$. The correlated samples $\mathbf{y}$ are then quantized using a uniform quantizer with step size Δ with q levels to obtain a vector $\mathbf{z} \in \mathbb{F}^n$. The samples of $\mathbf{z}$ are placed in independent packets and transmitted in the network where they undergo network coding operations represented by a matrix A. When A has full rank n, an estimate $\hat{x}$ of $\mathbf{x}$ can be obtained from the received packets $\mathbf{p}$ by inverting the network coding matrix A, which provides an estimate

$$\hat{\mathbf{z}} = Q^{-1}(A^{-1}\mathbf{p}) \qquad [8.29]$$

of $\mathbf{z}$, with Q^{-1} being the quantizer reconstruction function. This estimate is then used to obtain an estimate of $\mathbf{x}$

$$\hat{\mathbf{x}} = (F^T F)^{-1} F^T \hat{\mathbf{z}} \qquad [8.30]$$

When not enough packets are received, the coding matrix A cannot be inverted. Since no unique estimate of $\mathbf{x}$ can be inferred from received packets $\mathbf{p} \in \mathbb{F}^m$, we select the one with the minimal norm

$$\hat{\mathbf{x}} = \arg \min \mathbf{x}^T \mathbf{x} \qquad [8.31]$$

under the constraints

$$\begin{cases} \mathbf{p} = A\mathbf{z} \\ \mathbf{y} = F\mathbf{x} \\ \mathbf{z} \in \mathbb{F}^n \\ \mathbf{x} - \alpha\mathbf{z} - \beta \leqslant \Delta.\mathbf{1} \\ -\mathbf{x} + \alpha\mathbf{z} + \beta \leqslant \Delta.\mathbf{1} \end{cases} \qquad [8.32]$$

In [8.32], the first constraint allows us to take into account the received packets, the second, the fact that the vector to be estimated has been transformed using a frame. The third expresses the fact that each component of z belongs to $\{0, \ldots, q - 1\}$. The last two constraints allow us to take into account the bounded character of the quantization noise. This constrained optimization is difficult because it combines the real variables x and z with the components of z belonging to a Galois field. When the size of the Galois field q is prime, the network coding operations can be expressed in the ring of integers $\mathbb{Z}$ by introducing an additional vector $s \in \mathbb{Z}^m$ to express the first constraint as

$$\mathbf{p} = A\mathbf{z} + q\mathbf{s} \qquad\qquad [8.33]$$

The solution for [8.31] under constraints [8.32] where the first constraint is replaced by [8.33] requires the resolution of a mixed integer quadratic optimization problem. This type of problem can be modeled with AMPL and solved with CPLEX. Estimation complexity here is much higher than that of a redundancy introduction technique using a transformation in a Galois field. However, a part of the quantization noise can be suppressed when a high number of packets are received.

8.5.1.2. *Existing correlation*

In this case, we assume that the source(s) generate correlated data $\mathbf{x} \in \mathbb{R}^k$, assumed, for example, to be the realization of a Gaussian vector X of mean zero and a non-diagonal covariance matrix Σ. A typical scenario corresponds to several sensors dispersed geographically and taking correlated measures x_i, $i = 1, \ldots, k$. These measures are quantized to obtain the samples $z_i \in \mathbb{F}^k$ transmitted on the network where they are coded. The effect of network coding is represented by a coding matrix A, and a set of packets $\mathbf{p} = A\mathbf{z}$ is obtained at the collection point. A maximum *a posteriori* estimator of z from p is proposed in [IWA 12]

$$\widehat{\mathbf{z}} = \arg\max_{\mathbf{z}} P(\mathbf{z}|\mathbf{p}) \qquad\qquad [8.34]$$

The Gaussian probability distribution and the correlation between samples of x are exploited to obtain a new mixed quadratic optimization problem modeled with AMPL and solved with CPLEX. Robustness against losses increases with the correlation between the components of x.

8.5.2. *Joint source–network coding*

The aim of joint source–network coding in a network of sensors is to simultaneously collect and compress data in the network. This section examines the problem of joint source–network coding in presence of losses. We will consider a network of sensors described by a directed acyclical graph $\mathcal{G} = \{\mathcal{V}, \mathcal{E}\}$. Among the nodes of this graph, there are source nodes, $s \in \mathcal{S} \subset \mathcal{V}$, collector nodes $t \in \mathcal{T} \subset \mathcal{V}$, and intermediate nodes belonging to $\mathcal{E} \backslash (\mathcal{S} \cup \mathcal{T})$. Assume that there is a single collector node $|\mathcal{T}| = 1$, and that for every $s \in \mathcal{S}$, there is a path from s to t. The aim is to estimate in t the data $x_1, \dots, x_n$ with $x_i \in \mathbb{R}$ and $n = |\mathcal{S}|$, exploiting the correlation between x_is, in order to minimize the exchanges on the network. This problem is linked to robust network coding by the fact that the proposed collection scheme must eventually be robust to losses introduced by certain edges of the network.

The lossy coding method proposed in [LIU 12] assumes that the correlation between the x_is is known at t. The method consists of quantizing x_i at each source using a uniform quantizer on q levels. The quantized data are then sent on the network where they are network coded. The collector node t receives $m \leqslant n$ coded packets $y_1, \dots, y_m$ from which it has to estimate $x_1, \dots, x_n$. For this, it exploits the existing correlation between the x_is. It is possible to formally write a maximum *a posteriori* estimator of $\mathbf{x} = (x_1, \dots, x_n)$ from $\mathbf{y} = (y_1, \dots, y_m)$ as follows:

$$\widehat{\mathbf{x}}_{\text{MAP}} = \arg \max_{\mathbf{x}} p(\mathbf{x}|\mathbf{y}) \qquad [8.35]$$

which allows us to take into account all the information that t has. However, the evaluation complexity of $\hat{\mathbf{x}}_{\text{MAP}}$ is exponential in the number of sensors and, even for small networks, an exact implantation of [8.35] is not feasible. However, it is possible to represent the relationship between components of $\mathbf{x}$ and $\mathbf{y}$ using a bipartite graph [RIC 08], where the variable nodes are x_i and the *check* nodes are y_j, corresponding to the received packets, and the nodes z_k allow to account for the correlation between the components of $\mathbf{x}$, see Figure 8.4. Belief propagation algorithms [MAC 03, RIC 08] can then be employed to obtain an estimate of the *a posteriori* marginals $p(x_i|\mathbf{y})$ from which an *a posteriori* component-by-component approximation can be obtained. This estimation provides an approximate solution to [8.35]. This technique is effecient when the data generated by sensors are highly correlated. Moreover, matrix A representing the network coding operations (and which allows us to deduce the links between the x_i and y_j in the bipartite graph) should be sufficiently sparse to allow convergence of the belief propagation algorithm. The way in which network coding must be carried out to ensure that A has the correct properties remains, to our knowledge, an unresolved problem.

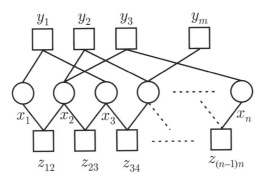

Figure 8.4. *Bipartite graph associated with the joint source–network coding problem*

8.6. Conclusion

This chapter has introduced several robust network coding techniques, which aim to cope with losses and errors introduced by links or nodes of the network. Coding techniques for coherent networks require the knowledge of the network structure and a centralized optimization of the way in which network coding is carried out. These techniques are well adapted to situations where the network structure is static. The main point of interest is that it is not necessary to introduce network coding coefficients into the packet headers passing through the network. Therefore, performance for this type of code is deterministic.

The previous techniques can be extended to a random network coding framework. This means that network coding is performed locally, in a distributed way, providing good adaptivity to variations in network topology. This type of code is, therefore, well adapted to mobile wireless networks. The disadvantage is that the performance of robust network code for coherent networks in this type of situation is described by random variables. It is not possible to guarantee a minimum distance for a robust network code. This requires transmitting the network coding coefficients in the packet headers, which leads to an increase in the amount of data on the links.

The subspace codes represent an interesting alternative to the previously mentioned techniques, more specifically in a non-coherent network. These techniques are particularly effective against deletions or against erroneous packets injected deliberately by some of the nodes of the network, see [NUT 08]. These tools can be associated with joint network–channel decoding techniques. The joint exploitation of the redundancy of the channel code and of the network code can significantly reduce the number of packets that would be considered as erroneous if only one separate process would have been carried out. The cost, however, is an increase in the decoding complexity at the receiver.

Source–network coding techniques have a particular significance for improving network robustness to loss. These techniques also allow us to have a more progressive improvement in the quality of decoded messages when the number of packets received by a destination increases. This can be interesting for transmitting multimedia contents, for example in pair-to-pair networks. See [CHE 07, MAR 09, MAG 09] for further details. Finally, network coding can be seen as a highly significant tool for joint source–network coding since it allows an effective collection of data across sensor networks.

8.7. Acknowledgments

This research has been partly supported by the DIM-LSC NC2! and SWAN projects. Michel Kieffer is partly supported by the Institut Universitaire de France.

8.8. Bibliography

[AHL 00] AHLSWEDE R., CAI N., LI S.Y.R., YEUNG R.W., "Network information flow", *IEEE Transactions on Information Theory*, vol. 46, no. 4, pp. 1204–1216, 2000.

[AHL 09] AHLSWEDE R., AYDINIAN H., "On error control codes for random network coding", *Proceedings of Workshop on Network Coding, Theory, and Applications*, Lausanne, Switzerland, pp. 68–73, June 2009.

[ALH 09] AL HABIAN G., GHRAYEB A., HASNA M., "Controlling error propagation in network–coded cooperative wireless networks", *Proceedings of IEEE International Conference on Communications*, Dresden, Germany, pp. 1–6, 2009.

[BAL 09] BALLI H., YAN X., ZHANG Z., "On randomized linear network codes and their error correction capabilities", *IEEE Transactions on Information Theory*, vol. 55, no. 7, pp. 3148–3160, 2009.

[BAO 06] BAO X., LI J., "A unified channel-network coding treatment for user cooperation in wireless ad hoc networks", *Proceedings of IEEE International Symposium on Information Theory*, Seattle, USA, pp. 202–206, 2006.

[BAR 06] BARROS J., SERVETTO S.D., "Network information flow with correlated sources", *IEEE Transactions on Information Theory*, vol. 52, no. 1, pp. 155–170, January 2006.

[BYR 08] BYRNE E., "Upper bounds for network error correcting codes", *Banff International Research Station (BIRS) Workshop: Applications of Matroid Theory and Combinatorial Optimization to Information and Coding Theory*, Banff, Canada, 2008.

[CAI 02] CAI N., YEUNG R., "Network coding and error correction", *Proceedings of IEEE Information Theory Workshop*, Bangalore, India, pp. 119–122, 2002.

[CAI 06] CAI N., YEUNG R.W., "Network error correction, part II: lower bounds", *Communications in Information and Systems*, vol. 6, no. 1, pp. 37–54, 2006.

[CAN 06] CANDES E., TAO T., "Near-optimal signal recovery from random projections: universal encoding strategies?", *IEEE Transactions on Information Theory*, vol. 52, no. 12, pp. 5406–5425, December 2006.

[CHE 09] CHEN N., GADOULEAU M., YAN Z., "Rank metric decoder architectures for noncoherent error control in random network coding", *IEEE Workshop on Signal Processing Systems*, Tampere, Finland, pp. 127–132, October 2009.

[CHE 07] CHENGUANG X., YINLONG X., CHENG Z., RUIZHE W., QINGSHAN W., "On network coding based multirate video streaming in directed networks", *Proceedings of IEEE International Performance, Computing, and Communications Conference*, New Orleans, USA, pp. 332–339, April 2007.

[CHO 03] CHOU P.A., WU Y., JAIN K., "Practical network coding", *Proceedings of 41st Annual Allerton Conference on Communication, Control, and Computing*, Monticello, USA, pp. 1–10, 2003.

[COL 09] COLEMAN E.M., ORDENTLICH E., "Joint source-channel coding for transmitting correlated sources over broadcast networks", *IEEE Transactions on Information Theory*, vol. 55, no. 8, pp. 3864–3868, 2009.

[CRI 05] CRISTESCU R., BEFERULL-LOZANO B., VETTERLI M., "Networked Slepian-Wolf: theory, algorithms, and scaling laws", *IEEE Transactions on Information Theory*, vol. 51, no. 12, pp. 4057–4073, December 2005.

[DON 06] DONOHO D.L., "Compressed sensing", *IEEE Transactions on Information Theory*, vol. 52, no. 4, pp. 1289–1306, 2006.

[DUH 09] DUHAMEL P., KIEFFER M., *Joint Source-Channel Decoding: A Cross-Layer Perspective with Applications in Video Broadcasting over Mobile and Wireless Networks*, EURASIP and Academic Press Series in Signal and Image Processing, Academic Press, Oxford, UK, 2009.

[EFF 03] EFFROS M., MEDARD M., HO T., RAY S., KARGER D., KOETTER R., "Linear network codes: a unified framework for source, channel and network coding", *DIMACS Workshop on Network Information Theory*, Newark, USA, 2003.

[ETZ 09] ETZION T., SILBERSTEIN N., "Error–correcting codes in projective spaces via rank–metric codes and ferrers diagrams", *IEEE Transactions on Information Theory*, vol. 55, no. 7, pp. 2909–2919, 2009.

[GAB 08] GABIDULIN E.M., BOSSERT M., "Codes for network coding", *Proceedings of IEEE International Symposium on Information Theory*, Toronto, Canada, pp. 867–870, July 2008.

[GAD 09] GADOULEAU M., YAN Z., "Bounds on covering codes with the rank metric", *IEEE Communications Letters*, vol. 13, no. 9, pp. 691–693, September 2009.

[GOY 98] GOYAL V.K., Beyond Traditional Transform Coding, PhD Thesis, University of California, Berkeley, USA, 1998.

[GOY 01] GOYAL V., KOVACEVIC J., KELNER J., "Quantized frame expansions with erasures", *Applied and Computational Harmonic Analysis*, vol. 10, no. 3, pp. 203–233, 2001.

[GUO 09] GUO Z., HUANG J., WANG B., CUI J.-H., ZHOU S., WILLETT P., "A practical joint network-channel coding scheme for reliable communication in wireless networks", *Proceedings of ACM International Symposium on Mobile Ad Hoc Networking and Computing*, New Orleans, USA, pp. 279–288, 2009.

[HAU 06a] HAUSL C., DUPRAZ P., "Joint network-channel coding for the multiple-access relay channel", *Proceedings of IEEE Communications Society Conference on Sensor and Ad Hoc Communications and Networks*, Reston, USA, pp. 817–822, 2006.

[HAU 06b] HAUSL C., HAGENAUER J., "Iterative network and channel decoding for the two-way relay channel", *Proceedings of IEEE International Conference on Communications*, Istanbul, Turkey, pp. 1568–1573, 2006.

[HO 03] HO T., KOETTER R., MEDARD M., KARGER D.R., EFFROS M., "The benefits of coding over routing in a randomized setting", *Proceedings of IEEE International Symposium on Information Theory*, Yokohama, Japan, June 2003.

[HO 04] HO T., MEDARD M., EFFROS M., KOETTER R., "Network coding for correlated sources", *Proceedings of IEEE Conference on Information Sciences and Systems*, Princeton, USA, pp. 1–6, 2004.

[HO 06] HO T., MÉDARD M., KOETTER R., KARGER D., EFFROS M., SHI J., LEONG B., "A random linear network coding approach to multicast", *IEEE Transactions on Information Theory*, vol. 52, no. 10, pp. 4413–4430, 2006.

[HOW 08] HOWARD L., FLIKKEMA P.G., "Integrated source-channel decoding for correlated data-gathering sensor networks", *Proceedings of IEEE Wireless Communications and Networks Conference*, Las Vegas, USA, April 2008.

[IWA 11] IWAZA L., KIEFFER M., LIBERTI L., AL AGHA K., "Joint decoding of multiple-description network-coded data", *Proceedings of International Symposium on Network Coding*, Beijing, China, July 2011.

[IWA 12] IWAZA L., KIEFFER M., AL AGHA K., MAP estimation of network-coded gaussian correlated sources, Technical report L2S 02/2012, 2012, http://www.lss.supelec.fr/~publi/TWljaGVsIEtJRUZGRVI=_Gaussian_Source_v5.pdf.

[JAG 05] JAGGI S., SANDERS P., CHOU P., EFFROS M., EGNER S., JAIN K., TOLHUIZEN L., "Polynomial time algorithms for multicast network code construction", *IEEE Transactions on Information Theory*, vol. 51, no. 6, pp. 1973–1982, 2005.

[JAG 07] JAGGI S., LANGBERG M., KATTI S., HO T., KATABI D., MEDARD M., "Resilient network coding in the presence of byzantine adversaries", *Proceedings of IEEE International Conference on Computer Communications*, Barcelona, Spain, pp. 616–624, 2007.

[KAT 05] KATTI S., KATABI D., HU W., RAHUL H., MÉDARD M., "The importance of being opportunistic: practical network coding for wireless environments", *Proceedings of 43rd Allerton Conference on Communication, Control, and Computing*, Monticello, USA, pp. 1–10, 2005.

[KAT 07a] KATTI S., GOLLAKOTA S., KATABI D., "Embracing wireless interference: analog network coding", *Proceedings of ACM Conference on Applications, Technologies, Architectures, and Protocols for Computer Communications*, Kyoto, Japan, pp. 397–408, 2007.

[KAT 07b] KATTI S., KATABI D., "MIXIT: the network meets the wireless channel", *Proceedings of ACM Workshop on Hot Topics in Networks*, Atlanta, USA, pp. 1–7, 2007.

[KHA 09] KHALEGHI A., KSCHISCHANG F., "Projective space codes for the injection metric", *Proceedings of 11th Canadian Workshop on Information Theory*, Ottawa, Canada, pp. 9–12, May 2009.

[KLI 07] KLIEWER J., DIKALIOTIS T., HO T., "On the performance of joint and separate channel and network coding in wireless fading networks", *Proceedings of IEEE Workshop on Information Theory for Wireless Networks*, Bergen, Norway, pp. 1–5, 2007.

[KOE 03] KOETTER R., MEDARD M., "An algebraic approach to network coding", *IEEE/ACM Transactions on Networking*, vol. 11, no. 5, pp. 782–795, 2003.

[KOE 08] KOETTER R., KSCHISCHANG F.R., "Coding for errors and erasures in random network coding", *IEEE Transactions on Information Theory*, vol. 54, no. 8, pp. 3579–3591, 2008.

[KOH 08] KOHNERT A., KURZ S., "Construction of large constant dimension codes with a prescribed minimum distance", *Lecture Notes in Computer Science*, Springer, pp. 31–42, 2008.

[LEE 07] LEE A., MEDARD M., HAIGH K., RUBEL P., "Minimum-cost subgraphs for joint distributed source and network coding", *IEEE Workshop on Network Coding, Theory and Applications*, San Diego, USA, pp. 1–4, January 2007.

[LI 03] LI S.-Y.R., YEUNG R.W., CAI N., "Linear network coding", *IEEE Transactions on Information Theory*, vol. 49, no. 2, pp. 371–381, 2003.

[LIU 12] LIU C., BASSI F., IWAZA L., KIEFFER M., Joint source-network coding for efficient data collection in wireless sensor networks, Technical report L2S 01/2012, 2012, http://www.lss.supelec.fr/~publi/TWljaGVsIEtJRUZGRVI=_ICASSP_Liu_2012.pdf.

[MAC 03] MACKAY D.J.C., *Information Theory, Inference, and Learning Algorithms*, Cambridge University Press, Cambridge, 2003.

[MAG 09] MAGLI E., FROSSARD P., "An overview of network coding for multimedia streaming", *Proceedings of International Conference on Multimedia and Expo*, New York, USA, pp. 1488–1491, June 2009.

[MAR 09] MARKOPOULOU A., SEFEROGLU H., "Network coding meets multimedia: opportunities and challenges", *IEEE MMTC E-Letter*, vol. 4, no. 1, pp. 12–15, 2009.

[MAT 07] MATSUMOTO R., "Construction algorithm for network error–correcting codes attaining the singleton bound", *IEICE Transactions on Fundamentals*, vol. E90-A, no. 9, pp. 1–7, 2007.

[NUT 08] NUTMAN L., LANGBERG M., "Adversarial models and resilient schemes for network coding", *Proceedings of IEEE International Symposium on Information Theory*, Toronto, Canada, pp. 171–175, July 2008.

[RIC 08] RICHARDSON T., URBANKE U., *Modern Coding Theory*, Cambridge University Press, 2008.

[SHI 08] SHINTRE S., KATTI S., JAGGI S., "Real and complex network codes: promises and challenges", *Proceedings of the 4th Workshop on Network Coding, Theory and Applications*, Hong Kong, China, pp. 1–6, 2008.

[SIL 08] SILVA D., KSCHICHANG F.R., KÖTTER R., "A rank-metric approach to error control in random network coding", *IEEE Transactions on Information Theory*, vol. 54, no. 9, pp. 3951–3967, 2008.

[SIL 09] SILVA D., KSCHISCHANG F., "On metrics for error correction in network coding", *IEEE Transactions on Information Theory*, vol. 55, no. 12, pp. 5479–5490, 2009.

[SLE 73] SLEPIAN D., WOLF J., "Noiseless coding of correlated information sources", *IEEE Transactions on Information Theory*, vol. 19, no. 4, pp. 471–480, 1973.

[THO 08] THOBABEN R., "Joint network/channel coding for multi-user hybrid-ARQ", *Proceedings of International ITG Conference on Source and Channel Coding*, Ulm, Germany, pp. 1–6, 2008.

[WU 05] WU Y., STANKOVIC V., XIONG Z., KUNG S.Y., "On practical design for joint distributed source and network coding", *Proceedings of the 1st Workshop on Network Coding, Theory, and Applications*, Riva del Garda, Italy, pp. 1–5, April 2005.

[WYN 76] WYNER A., ZIV J., "The rate-distortion function for source coding with side information at the decoder", *IEEE Transactions on Information Theory*, vol. 22, no. 1, pp. 1–10, 1976.

[XIA 07] XIAO L., FUJA T., KLIEWER J., COSTELLO D., "A network coding approach to cooperative diversity", *IEEE Transactions on Information Theory*, vol. 53, no. 10, pp. 3714–3722, 2007.

[XIA 08] XIA S., FU F., "Johnson type bounds on constant dimension codes", *Designs, Codes, and Cryptography*, vol. 50, no. 2, pp. 163–172, 2008.

[YAN 08] YANG S., YEUNG R., ZHANG Z., "Weight properties of network codes", *IEEE Transactions on Information Theory*, vol. 19, no. 4, pp. 371–383, 2008.

[YAN 11] YANG S., YEUNG R., NGAI C., "Refined coding bounds and code constructions for coherent network error correction", *IEEE Transactions on Information Theory*, vol. 57, no. 3, pp. 1409–1424, 2011.

[YEU 05] YEUNG R.W., LI S.-Y.R., CAI N., ZHANG Z., "Network coding theory", *Foundations and Trends in Communications and Information Theory*, vol. 2, nos. 4–5, pp. 241–381, 2005.

[YEU 06] YEUNG R.W., CAI N., "Network error correction, part I: basic concepts and upper bounds", *Communications in Information and Systems*, vol. 6, no. 1, pp. 19–36, 2006.

[YIN 09] YING L., SONG G., WANG L., "Design of joint network-low density parity check codes based on the EXIT charts", *IEEE Communications Letters*, vol. 13, no. 8, pp. 600–602, 2009.

[ZHA 08] ZHANG Z., "Linear network error correction codes in packet networks", *IEEE Transactions on Information Theory*, vol. 54, no. 1, pp. 209–218, 2008.

[ZHA 11] ZHANG Z., "Theory and applications of network error correction coding", *Proceedings of the IEEE*, vol. 99, no. 3, pp. 406–420, 2011.

Chapter 9

Flow Models and Optimization for Network Coding

9.1. Introduction

The notion of network coding has been introduced in [AHL 00] as a means of coding information flow in a network to maximize useful throughput. The main argument motivating this new coding technique was finding a means of exceeding the "classic" limitation of throughput in a multicast network and to reach maximal theoretical throughput, which would be attained only if the source transmits to one destination at a time.

In telecommunications networks, transferred information is, for the most part, split into smaller pieces (packets in IP) with each part then being routed independently. However, it is often necessary to model the circulation of all the packets in one connection as a more or less continuous flow of information (sometimes referred to as "fluid limit"). This simplified and,

Chapter written by Eric GOURDIN and Jeremiah EDWARDS.

to a certain extent, macroscopic view of information flow in telecommunications networks allows the use of a powerful analysis and modeling tool, namely graph theory (within the inherent limitations of any model), with the nodes and interconnecting links naturally becoming graph nodes and links, and the information flow being modeled as flows.

For more than half a century, graph theory focused on flow problems, and the highly abundant amount of literature on the subject allows us to very quickly identify and characterize a flow problem. Graph theory has developed as an independent field of optimization, but numerous results can also be analyzed using linear programming models.

In this chapter, we demonstrate how the network coding paradigm can be analyzed using flow models in graphs and how some problems can be solved using optimization methods. The advantage of these transformations is that we can access a number of results from graph theory relating to flow, reinterpret network coding with these tools, and offer different approaches to solve some problems, mainly related to network coding throughput maximization.

A number of flow models have already been proposed and used in various contexts. In a multicast network without coding, the problem of finding a minimal cost routing is similar to the Steiner tree problem, a classical NP-complete problem [RAM 96]. Somehow surprisingly, the network coding proposed in [AHL 00] and developed in [KOE 03] and [HO 06] adds a mechanism to the network, which simplifies the optimization problems. Lun *et al.* [LUN 04] have proposed optimization models that find the optimum paths in a network with coding. This approach is also adopted and generalized in [LUN 05] and resolution methods are detailed in [LUN 06]. These models have highly important implications for managing modern networks, for example in content distribution [HUA 09].

9.2. Some reminders on flow problems in graphs

For the terminology and the way of introducing and defining flow, we will use the classic notations described in A. Shrijver's *Combinatorial Optimization* [SCH 03].

A graph is a set V of nodes (or vertices) and a set of relations between these nodes. Directed graphs (or digraphs) are defined by sets of arcs $(i, j) \in A$, whereas undirected graphs are defined by sets of edges $\{i, j\} = \{j, i\} \in E$. Since undirected graphs are, to a certain extent, special cases of directed graphs, we will hence focus our attention on the more general case. We will therefore consider a directed graph $G = (V, A)$. Figure 9.1 provides an example of such a graph with $V = \{a, b, c, d, e\}$ and $A = \{(a, b), (a, d), (b, d), (c, a), (d, a), (d, d), (d, e)\}$. We can see that between the nodes a and d, there is an arc in both directions. Therefore, in some cases we can assimilate these two arcs to a single edge linking the two nodes. The arc (d, d) is called a loop. Even if these loops can be useful in certain contexts, we will suppose here that the graphs do not have loops. We denote $n = |V|$ and $m = |A|$ (for the example in Figure 9.1, $n = 5$ and $m = 7$).

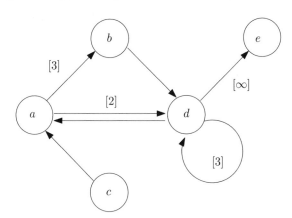

Figure 9.1. *An example of a (directed) graph with five nodes and seven arcs*

Given a source node $s \in V$ and a destination node $t \in V$, a flow between s and t, or s–t flow, is a function $f : A \to \mathbb{R}$ such that

$$f(a) \geq 0, \qquad\qquad a \in A, \qquad\qquad\qquad\text{[9.1]}$$

$$f(\delta^-(v)) = f(\delta^+(v)), \quad v \in V \setminus \{s, t\}. \qquad\qquad\text{[9.2]}$$

We denote $\delta^-(v)$ ($\delta^+(v)$, respectively) as the set of incoming arcs (or outgoing arcs, respectively) of the node $v \in V$. On the graph in Figure 9.1, we have, for example, $\delta^-(d) = \{(a, d), (b, d), (d, d)\}$ and $\delta^+(a) = \{(a, b), (b, d)\}$. In [9.2], we also use the compact notation $f(A')$ for $\sum_{a \in A'} f(a)$, where A' is a subset of arcs. As such, the constraints [9.2] (known as "flow conservation constraints") stipulate that in every node (except the source and destination), the total incoming flow is equal to the outgoing flow. In other words, there is a conservation of the flow.

The value of the s–t flow f, denoted by $v(f)$, is the amount of flow exiting the source s and reaching the destination t

$$v(f) = f(\delta^+(s)) - f(\delta^-(s)) = f(\delta^-(t)) - f(\delta^+(t)). \quad\text{[9.3]}$$

The notion of flow in a graph has little meaning if there are no limitations on the amount of flow that each arc can carry. We will use $C(a) > 0$ to denote the capacity of the arc $a \in A$. In the figures, the arcs' capacity is indicated between brackets with the convention that capacity is unitary unless otherwise indicated. On the example in Figure 9.1, the arcs $(a, b), (a, d)$, and (d, d) have the capacity of 3, 2, and 3, respectively. The arc (d, e) has an "infinite" capacity, therefore not limiting the flow. The other arcs have a unitary capacity: $C_{(b,d)} = C_{(c,a)} = C_{(d,a)} = 1$.

A flow f is feasible if it satisfies all the capacity constraints:

$$f(a) \leq C(a), \; a \in A. \qquad\qquad\qquad\qquad\text{[9.4]}$$

A problem that naturally arises in this context is knowing the maximal flow quantity that can be routed in the graph between the source and the destination. This is the well-known *maximum-flow* problem. If we denote $\mathcal{F}_{st}$ as the set of all the feasible s–t flows, then the maximum-flow (or max-flow) problem can be simply written as

$$\max_{f \in \mathcal{F}_{st}} v(f). \qquad\qquad\qquad [9.5]$$

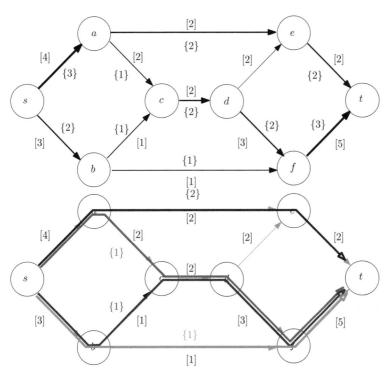

Figure 9.2. *An example of max-flow with a value of 5 between s and t (above: the flow, between brackets, on each arc, below: equivalent flow on each of the four paths)*

We will write v_{st}^* as the flow's maximum value. Figure 9.2 provides an example of maximum flow. The values between brackets $\{f_a\}$ are the values of the flow on each arc (which has a value of 0 unless otherwise indicated). We can see that an s–t flow can be broken down into a set of flows on the paths between s and t. On the lower part of Figure 9.2, the maximum flow of 5 units is broken down on four paths as follows:

- 2 units on $s - a - e - t$,
- 1 unit on $s - a - c - d - f - t$,
- 1 unit on $s - b - c - d - f - t$,
- 1 unit on $s - b - f - t$.

One of the fundamental results related to this problem is the famous max-flow/min-cut theorem. We call a cut between s and t, or s–t cut, as any set of arcs, written as $\delta^+(S)$ or $A(S, \overline{S})$, induced by a partition of V into two subsets S and $\overline{S}$ such that $s \in S$ and $t \in \overline{S}$.

$$\delta^+(S) = A(S, \overline{S}) = \{(i, j) \in A : i \in S, j \in \overline{S}\}. \qquad [9.6]$$

If we denote $C(S, \overline{S})$ as the ability of the cut, that is $\sum_{a \in A(S, \overline{S})} C(a)$, then the minimum-cut problem (or min-cut) is written as

$$\min_{S \subset V} \{C(S, \overline{S}) : s \in S, t \in \overline{S}\}. \qquad [9.7]$$

Figure 9.3 shows three possible s–t cuts, induced by $\{s\}, \{s, a, b, c\}$, and $\{s, a, b, c, d\}$, with the respective values 7, 5, and 8.

We denote c^*_{st} as the value of the minimum s–t cut. It is fairly easy to see that the value of every s–t cut is an upper bound for every feasible s–t flow. We say that there is a weak duality between the two problems (this type of property can frequently be observed). However, the result, called strong duality, that stipulates that the two optimal values are equal, is much rarer.

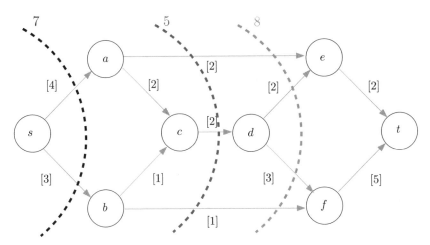

Figure 9.3. *Examples of three s–t cuts*

THEOREM   9.1.–   *(Max-flow/min-cut)*   *[DAN 56,   FOR 54,*
FOR 56]: $v_{st}^* = c_{st}^*$.

One of the main points of interest in this theorem is that we can "qualify" solutions of max-flow and min-cut problems. In particular, if we find a solution to each of the problems where the values are equal, then we know that the solutions are optimal. For the example in Figures 9.2 and 9.3, we have highlighted a flow and a cut, both with a value of 5. Therefore, these are the optimal solutions.

Unfortunately, this strong duality result is not generalized when we consider several flows (e.g. for multicommodity-flow problems or concurrent flows) [GAR 93, LEI 99, SHA 91] or other types of information transfers.

When we want to focus on protocols or routing mechanisms in telecommunications networks, there are often specificities that can be translated into flow variants. In particular, we can cite:

1) *Uni- or multicast*: the nodes (routers) can either transmit the flow or replicate it on several exit interfaces.

2) *Coding, compression, or none*: the nodes (routers) can code or combine several flows toward a single exit interface.

3) *Mono- or multi-routing*: the same flow (between s and t) can be routed on a single path or split on several paths.

4) *Redundancy or none*: for the purpose of survivability, flows can be partially or entirely duplicated locally or end to end so that in case of failure or attack, the whole flow can still reach its destination.

There are numerous flow models, some of which enable us to model the aforementioned variant. Here we will focus exclusively on examples related to network coding including points 1 (multicast) and 2 (network coding). Note that the notion of network coding, such as that studied in the literature, often involves the notion of multicast transfer.

9.3. Flow models for multicast traffic

With some telecommunications services (live TV, teleconferencing, etc.), several clients may request the same content simultaneously. With "standard" (unicast) protocols, we then need to transfer the same content several times on a single link (Figure 9.4 (a)), leading to a waste of network resources. Multicast protocols have therefore been developed to avoid this problem. With a multicast protocol, some of the network's nodes have the ability to replicate the traffic flow that they receive on several exit interfaces and therefore toward several different destinations (Figure 9.4 (b)).

Unfortunately, the transfer of flows obtained by multicast protocols are more difficult to model in the form of flows. We can see that with a multicast protocol, the flow conservation constraint is no longer valid (since, by definition, a node can transmit more than it receives):

$$f(\delta^-(v)) \leq f(\delta^+(v)), v \in V \setminus \{s, t\}.$$ [9.8]

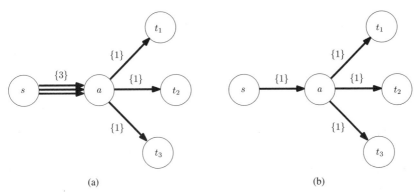

(a)   (b)

Figure 9.4. *Simultaneous transfer of three flows: (a) without multicast, (b) with multicast*

Although this simplification (relaxation) of the constraint may seem trivial, it has important consequences for the model's validity. In particular, we can lose connectivity between the source and the destination (the flow does not arrive at its destinations) and can also generate useless flows.

Multicast flow models have been extensively studied in different contexts [GAU 02, SAH 00, YOU 04]. One of the most studied problems consists of determining a minimal-cost multicast tree. To do so, we allocate a weight w_a to each arc $a \in A$. This problem is similar to a well-known problem in undirected graphs called the *Steiner tree* [BEA 84, GOE 93]. While this is a difficult problem (NP-complete), effective resolution techniques based on flow models have been proposed to solve it. These models can be adapted to the context of directed graphs and as such enable us to treat the problem of a minimum-cost multicast tree.

$$\min \sum_{a \in A} w_a y_a, \tag{9.9}$$

s.c.: $f^t(\delta^-(v)) - f^t(\delta^+(v)) = 0, \quad t \in T, v \in V \setminus \{s, t\}, \tag{9.10}$

$\qquad f^t(\delta^-(s)) - f^t(\delta^+(s)) = -1, \; t \in T, \tag{9.11}$

$\qquad f^t(\delta^-(t)) - f^t(\delta^+(t)) = +1, \; t \in T, \tag{9.12}$

$\qquad f^t(a) \le y_a, \qquad\qquad\qquad t \in T, a \in A, \tag{9.13}$

$\qquad f^t(a) \ge 0, y_a \in \{0,1\}, \qquad t \in T, a \in A. \tag{9.14}$

In this model, we associate a flow f^t with each terminal $t \in T$. This flow therefore moves between s and t (flow conservation constraints [9.10], [9.11], and [9.12]). Thanks to the constraints [9.13], the binary variables y_a take the value 1 if (and only if) there is at least one flow that passes through the arc a. As such, the variables y_a allow us to identify the multicast tree. Note that it is the minimization of the objective function [9.9] that implies that the optimal solution to the problem will be a tree (and not a meshed subgraph), if such a solution exists. In particular, the choice of objective function means that the constraints [9.13] imply that

$$y_a = \min_{t \in T} f^t(a), \; a \in A. \tag{9.15}$$

The functioning principle of this model is illustrated in Figure 9.5: on the graph on the left-hand side, three flows are sent individually from the source s toward each of the three destinations t_1, t_2, and t_3. The movement of at least one flow on an arc leads to the activation of this arc in a multicast tree, which appears on the graph on the right. We see that the formed tree reaches all the terminals and uses some of the other nodes (sometimes also known as *Steiner nodes*), but not all. This tree is therefore not a spanning tree (covering all the nodes) that makes the problem difficult.

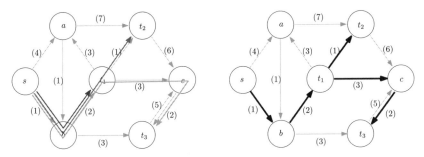

Figure 9.5. *Multicast tree with a minimum cost (Steiner). The costs of the arcs are indicated between brackets. On the left, the flow variables f^t with $t = 1, 2, 3$; on the right, the variables y_a (the arcs in bold)*

There are a number of other means of modeling this problem. In particular, we can see that in models [9.9]–[9.14], several conservative flows are used to generate the tree structure. As mentioned earlier in [9.8], we can also model the non-conservative flow from the source toward the terminal:

$$\min \quad \sum_{a \in A} w(a) f(a), \tag{9.16}$$

under the constraints: $f(\delta^-(v)) \le f(\delta^+(v)), \quad v \in V \setminus \{s, t\},$

$$\tag{9.17}$$

$$f(\delta^+(s)) \ge 1, \tag{9.18}$$

$$f(\delta^-(t)) = 1, \qquad t \in T, \tag{9.19}$$

$$f(a) \in \{0, 1\}, \qquad a \in A. \tag{9.20}$$

Constraint [9.18] stipulates that the flow must necessarily exit the source s and the constraint [9.19] stipulates that a flow unit must arrive at each terminal (if there is no flow unit arriving at a terminal, the formed structure is no longer a tree). Figure 9.6 shows an example of an optimal solution for models [9.16]–[9.20]. We can verify that all the constraints are satisfied. However, the produced solution is not correct from a telecommunications point of view, since the source s is not connected to the three terminals.

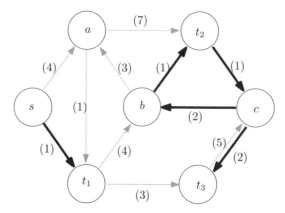

Figure 9.6. *An optimal but not connected solution for models [9.16]–[9.20]*

One means of overcoming this inconvenience is to introduce the so-called connectivity constraints:

$$f(\delta^+(S)) \geq 1, \ S \subset V, s \in S, S \cap T \neq T. \tag{9.21}$$

This constraint means that for every subset of nodes S containing the source but not all the terminals, there has to be at least one exit flow unit. We can see in the example in Figure 9.6 that this is not the case, for example for the subset $S = \{s, t_1\}$.

We can see that there is a very high (exponential) number of such constraints. It is therefore not conceivable to solve an instance of reasonable size by enumerating all the constraints. However, it is possible to solve problems in practice by generating only a very restricted number of connectivity constraints (typically only those that are not verified by a current solution). Attentive readers would have noticed that these connectivity constraints are alone sufficient for guaranteeing a solution's validity. However, adding the constraints [9.17]–[9.20] allows us to reinforce the model (the details of which we will not examine here), thus improving the resolution of the problem.

9.4. Flow models for network coding

The principle of network coding has been extensively explained in this book. With our instance of "flow" we can say that network coding allows the compression of several flows into one single flow, in addition to multicast features. We therefore see that the example in Figure 9.7 mirrors the example in Figure 9.4: in (a), the three flow units A, B, and C are passing together on the arc (a, t), but using three capacity units while in (b), the three flows are coded into a single flow, which only uses one capacity unit.

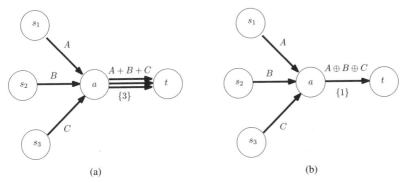

Figure 9.7. *Transfer of three aggregated flows on one link, (a) without network coding, (b) with network coding*

In [AHL 00], the authors show that throughput that can be reached between a source and a set of terminals is "maximal" if we use network coding. The "classic" example is a "butterfly network" (see Figure 9.8): in this graph there is a source s and two terminals t_1 and t_2. We consider that all the arcs have a capacity of one and that we want to maximize the amount of information that we can send toward the two destinations.

According to what has been previously said, we see that the max-flow between s and t_1 has a value of 2 (e.g. by routing one unit on the path $s - a - t_1$ and a unit on the path $s - b - c - d - t_1$, as indicated on the figure to the right). Equally, by extension, we can route two flow units toward t_2. However, we see that, while without network coding we cannot carry out two operations at the same time because, when the two units are routed toward t_1, there is not enough capacity available to route the request that passes on $s - a - t_1$ toward t_2.

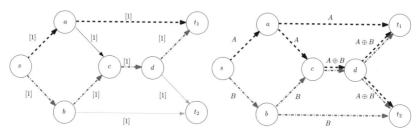

Figure 9.8. *A butterfly network*

The main result in [AHL 00] indicates that by using network coding it is possible to simultaneously send two flow units to t_1 and t_2, as illustrated on the figure to the right. More specifically, if we have a source s and a terminal $t_1, \ldots, t_k$, and if we write $c^*_{st_\ell}$ as the value of s–t_ℓ min-cut, then the flow that we can route simultaneously between s and $T = \{t_1, \ldots, t_k\}$ with network coding is

$$C^*_{sT} = \min_{\ell \in T} c^*_{st_\ell}. \qquad [9.22]$$

It therefore follows that it is easy to calculate C^*_{sT}: it is sufficient to calculate all the s–t_ℓ max-flow values and to keep that of the minimum value.

The calculation of k maximum flows therefore allows, on the one hand, to find the maximum throughput possible with a given coding scheme and provides, on the other hand, individual routing (means of transmitting the flow) toward each destination. The "superposition" of these individual routings generally allows us to identify a means of simultaneously transmitting information flows fairly easily. On the example in Figure 9.9, the maximum flow toward each of the three terminals t_1, t_2, and t_3 has a value of 3 (we again suppose that all the arcs have a capacity normalized at 1). The max-flows toward each destination are indicated on the figure to the right (there are three disjoint paths to reach t_1 and t_3 and five possible paths to reach t_2). It is therefore sufficient to associate a single stream with each path from s on the *same exit arc*. As such, we associate the stream "A" to the three arc exiting s on the arc $s-v_1$, the stream "B" to the three flows on the arc $s-v_2$, and the stream "C" to the three flows on the arc $s-v_3$. Increasingly, we can extend these streams by reducing them at specific nodes (multicast nodes) and coding them at other nodes (coding nodes). For example, the node v_2 reduces the stream "B" toward v_4, v_5, and t_2 and the node v_4 codes streams "A" and "B" into "$A \oplus B$".

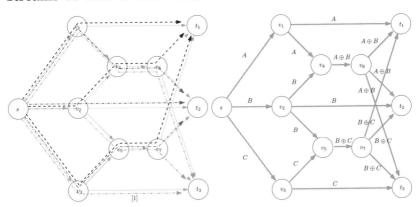

Figure 9.9. *A "double" butterfly network with three terminals*

A difficulty remains when there are several solutions to different maximum-flow problems (i.e. several ways of routing the flows to reach the same maximum-flow value). In this case, depending on the individual routing solutions chosen, the global solution can be significantly different. As such, by slightly modifying the previous example, we end up with the example shown in Figure 9.10, where the two arcs, (s, v_1) and (s, v_3), have a capacity of 2 (all the other arcs still having a capacity of 1). The previous solution is still valid and shows that we can transmit three streams A, B and C with network coding. However, there is another solution (figure to the right) where the three streams are transmitted but without resorting to network coding (only "standard" multicast). This therefore raises the question whether we should rely on network coding or not. The use of a linear model allows us to address this question. Let us suppose that we have identified the maximum throughput achievable (by solving k max-flow problems), which has a value of d_{max}. To choose from the equivalent solutions, one of those that minimizes routing "cost", we can therefore solve the following linear problem [LUN 05]:

$$\min \sum_{a \in A} w(a)y(a),$$
[9.23]

$$\text{s.c.: } f^t(\delta^-(v)) - f^t(\delta^+(v)) = 0, \qquad t \in T, v \in V \setminus \{s, t\},$$
[9.24]

$$f^t(\delta^-(s)) - f^t(\delta^+(s)) = -d_{max}, t \in T,$$
[9.25]

$$f^t(\delta^-(t)) - f^t(\delta^+(t)) = +d_{max}, t \in T,$$
[9.26]

$$f^t(a) \leq y(a) \leq C(a), \qquad t \in T, a \in A,$$
[9.27]

$$f^t(a), y(a) \geq 0, \qquad t \in T, a \in A.$$
[9.28]

This problem is very similar to [9.9]–[9.14], except that here the flows are not unitary and we do not want to transmit them on one tree. If we want to verify whether there is a solution to this problem that does not use network coding (but only multicast diffusion), we have to add the constraints[1]:

$$y(\delta^-(v)) \leq y(\delta^+(v)), \ v \in V \setminus T. \qquad [9.29]$$

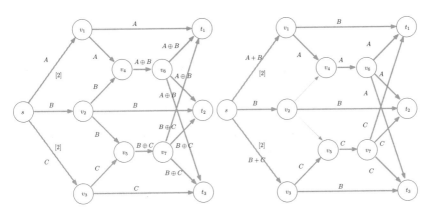

Figure 9.10. *Solutions with and without coding*

Network coding occurs only when the sum of entry flows is higher than the sum of exit flows (in which case coding must be used to "compact" the flows). If there is no solution without network coding, the problem above is therefore not feasible. Similarly, if we do not want certain nodes (in a set V_1) to be able to diffuse flows in multicast, we have to apply the following constraint:

$$y(\delta^-(v)) \geq y(\delta^+(v)), \ v \in V_1. \qquad [9.30]$$

If we want to prevent other nodes (in a set V_2) performing any function other than the flow transfer (neither multicast nor coding), we have to apply the following constraint:

$$y(\delta^-(v)) = y(\delta^+(v)), \ v \in V_2. \qquad [9.31]$$

---

1 In practice, this type of constraint is often sufficient for the aim in question even if, strictly speaking, the models proposed here are in fact approximations (relaxations).

The importance of this type of approach is that we can easily introduce a number of additional constraints into the model. For example, capacity constraints at the level of interfaces or processing imposed by network equipment can be included in models without necessarily making the solution more difficult. Let us suppose that, for each node $v \in V$, the number of interfaces at entry and exit is limited to Δ_{max}^- and Δ_{max}^+ respectively, then it is sufficient to add the constraints

$$y(\delta^-(v)) \leq \Delta_{max}^-, \ y(\delta^+(v)) \leq \Delta_{max}^+, \ v \in V. \qquad [9.32]$$

If we want to prevent some flows being routed on specific arcs, we have to fix the corresponding flows at 0 or restrict them. For example, if we require that the flow at the destination of the terminal t_ℓ must not pass by the arcs of the subset $A_\ell \subset A$, it suffices to add the following constraint:

$$f^{t_\ell}(a) = 0, \ a \in A_\ell, \qquad [9.33]$$

which essentially amounts to removing certain variables from the model.

As we saw earlier, the introduction of certain constraints can also complicate the model if, for example, we limit or reserve special treatment for nodes in the network that effectively carries out coding and/or multicast diffusion. This type of constraint can be modeled by introducing new binary or integer variables. In the example in Figure 9.11, we have deliberately limited the number of nodes that could carry out coding at $p = 1$ (knowing that in the optimal solution, two are required to have the throughput for three flow units toward each terminal). Here, only the node v_4 carries out coding of the flow A with half of the flow B. The node v_5 sends the two half-flows $B/2$ and $C/2$ together (without coding) toward v_7 (which is possible given that the link easily has a capacity of 1). The result of this limitation in terms of decoding in the network is that throughput passes from 2 to 2.5. After decoding, the terminal t_1 receives $A + B + C/2$ (it has lost half of the flux C). Similarly, the two other terminals lose half of one of the entry flows. As an optimal solution to the adequate linear problem, we know that we can achieve higher throughput with a single coding node.

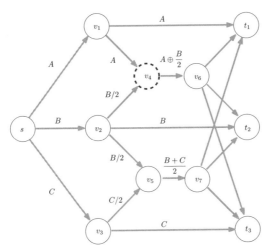

Figure 9.11. *Solution with limitation at a single coding node*

By using "arc-path" models (where variables are associated with paths rather than flows on the arcs), we can generally integrate constraints on the length of the paths fairly easily. In [LUN 04, LUN 05], the authors propose a model for finding routing (compatible with a network coding scheme) that carries out a particular (fixed) throughput and that minimizes a function representing the routing "cost". The authors provide a general form for this cost function that can be either linear or convex (without making the problem much more difficult to solve). We can compare these routing cost models with the various objective functions that were proposed at the beginning to model delays in TCP/IP networks. An example of this is the function proposed by Kleinrock [KLE 64]:

$$F_2 = \sum_{a \in A} \frac{1}{C(a) - y(a)} \qquad [9.34]$$

which has subsequently been generalized by Mo and Walrand [MO 00], by introducing the parameter α:

$$F_\alpha = \frac{1}{1 - \alpha} \sum_{a \in A} (C(a) - y(a))^{1-\alpha} \qquad [9.35]$$

which are convex functions. On the basis of the initial modeling objective, this family of functions allows us to introduce a notion of fairness to the allocation of network resources to different flows. It is for this reason that these functions have also been used extensively in the "traffic engineering" context to find both optimal and fair means of routing traffic. We therefore include the F_α function in one of the earlier described models to calculate fair routing in a network where the nodes can code flows.

These examples provide only a brief insight into the potential of optimization problems (linear or convex, continuous or integer numbers). Many constraints can be integrated more or less easily into these models. Even if the resulting problems are theoretically non-tractable, the construction of a good model is still of great help in the research of approximate or even optimal solutions.

9.5. Conclusion

In this chapter we have examined several flow and multicommodity-flow problems in graphs and have shown that they can be modeled as either linear or convex problems, with continuous variables or integers. The vast amount of literature and numerous results already obtained for these problems allow us to treat them, for the most part, very effectively. This results in flow models, reducing the limitations inherent to each model (a certain abstraction of reality, an often important resolution complexity, etc.), which are highly powerful tools for analyzing traffic flow problems in networks generally and the impact and performance of network coding in particular.

9.6. Bibliography

[AHL 00] AHLSWEDE R., CAI N., LI S.-Y., YEUNG R., "Network information flow", *IEEE Transactions on Information Theory*, vol. 46, no. 4, pp. 1204–1216, July 2000.

[BEA 84] BEASLEY J., "An algorithm for the Steiner problem in graphs", *Networks*, vol. 14, pp. 147–159, 1984.

[DAN 56] DANTZIG G., FULKERSON D., "On the max-flow min-cut theorem of networks", in KUHN H., TUCKER A., (eds), *Linear Inequalities and Related Systems*, Princeton University Press, Princeton, USA, 1956.

[FOR 54] FORD L., FULKERSON D., Maximum flow through a network, Research Memorandum RM-1400, Report, The RAND Corporation, Santa Monica, USA, 1954.

[FOR 56] FORD L., FULKERSON D., "Maximum flow through a network", *Canadian Journal of Mathematics*, vol. 8, pp. 399–404, 1956.

[GAR 93] GARG N., VAZIRANI M.Y.V., "Approximate max-flow min-(multi)cut theorems and their applications", *STOC'93*, pp. 698–707, 1993.

[GAU 02] GAU R.-H., HAAS Z., KRISHNAMACHARI B., "On multicast flow control for heterogeneous receivers", *IEEE/ACM Transactions on Networking*, vol. 10, no. 1, pp. 86–101, February 2002.

[GOE 93] GOEMANS M., MYUNG Y., "A catalog of Steiner tree formulations", *Networks*, vol. 23, pp. 19–28, 1993.

[HO 06] HO T., MÉDARD M., KOETTER R., KARGER D., EFFROS M., SHI J., LEONG B., "A random linear network coding approach to multicast", *IEEE Transactions on Information Theory*, vol. 52, no. 10, pp. 4413–4430, October 2006.

[HUA 09] HUANG S., RAMAMOORTHY A., MÉDARD M., "Minimum cost content distribution using network coding: replication vs. coding at the source nodes", *CoRR*, vol. abs/0910.2263, 2009.

[KLE 64] KLEINROCK L., *Communication Nets: Stochastic Message Flow and Delay*, McGraw-Hill, New York, USA, 1964.

[KOE 03] KOETTER R., MÉDARD M., "An algebraic approach to network coding", *IEEE/ACM Transactions on Networking*, vol. 11, no. 5, pp. 782–795, October 2003.

[LEI 99] LEIGHTON T., RAO S., "Multicommodity max-flow min-cut theorems and their use in designing approximation algorithms", *JACM*, vol. 46, pp. 215–245, 1999.

[LUN 04] LUN D.S., MÉDARD M., HO T., KOETTER R., "Network coding with a cost criterion", *Proceedings of 2004 International Symposium on Information Theory and its Applications (ISITA 2004)*, pp. 1232–1237, 2004.

[LUN 05] LUN D.S., RATNAKAR N., KOETTER R., MÉDARD M., AHMED E., LEE H., "Achieving minimum-cost multicast: A decentralized approach based on network coding", *Proceedings of IEEE Infocom*, pp. 1607–1617, 2005.

[LUN 06] LUN D.S., RATNAKAR N., MÉDARD M., KOETTER R., KARGER D.R., HO T., AHMED E., ZHAO F., "Minimum-cost multicast over coded packet networks", *IEEE/ACM Transactions on Networking*, vol. 14, no. SI, pp. 2608–2623, 2006.

[MO 00] MO J., WALRAND J., "Fair end-to-end window-based congestion control", *IEEE/ACM Transactions on Networking*, vol. 8, no. 5, pp. 556–567, 2000.

[RAM 96] RAMANATHAN S., "Multicast tree generation in networks with asymmetric links", *IEEE/ACM Transactions on Networking*, vol. 4, no. 4, pp. 558–568, August 1996.

[SAH 00] SAHASRABUDDHE L., MUKHERJEE B., "Multicast routing algorithms and protocols: A tutorial", *IEEE Networks*, vol. 14, pp. 90–102, 2000.

[SCH 03] SCHRIJVER A., *Combinatorial Optimization – Polyhedra and Efficiency*, Springer-Verlag, Berlin Heidelberg, 2003.

[SHA 91] SHAHROKHI F., MATULA D., "On solving large maximum concurrent flow problems", *Journal of the ACM*, vol. 37, pp. 318–334, 1991.

[YOU 04] YOUSEFIZADEH H., FAZEL F., JAFARKHANI H., "Hybrid unicast and multicast flow control: A linear optimization approach", in MAMMERI Z., LORENZ P. (eds) *High Speed Networks and Multimedia Communications*, Lecture Notes in Computer Science, pp. 369–380, Springer, Berlin Heidelberg, Germany, 2004.

List of Authors

Khaldoun AL AGHA
LRI
Paris-Sud University
France

Anya APAVATJUT
CITI-INSA Lyon
INRIA
France

Youghourta BENFATTOUM
LRI
Paris-Sud University
France

Antoine O. BERTHET
Supélec
Gif-sur-Yvette
France

Marco DI RENZO
L2S
CNRS-Supélec
Paris-Sud University
France

Jeremiah EDWARDS
Orange Labs
Issy-les-Moulineaux
France

Maximilien GADOULEAU
Durham University
UK

Jean-Marie GORCE
CITI-INSA Lyon
INRIA
France

Eric GOURDIN
Orange Labs
Issy-les-Moulineaux
France

Claire GOURSAUD
CITI-INSA Lyon
INRIA
France

Atoosa HATEFI
Supélec
Gif-sur-Yvette
France

Lana IWAZA
L2S
CNRS-Supélec
Paris-Sud University
France

Katia JAFFRÈS-RUNSER
IRIT
INPT-ENSEEIHT
University of Toulouse
France

Nour KADI
LIUM
University of Maine
Le Mans
France

Michel KIEFFER
L2S
CNRS-Supélec
Paris-Sud University
France

Cédric LAURADOUX
INRIA Rhône-Alpes
Grenoble
France

Steven MARTIN
LRI
Paris-Sud University
France

Marine MINIER
CITI-INSA Lyon
INRIA
France

Raphaël VISOZ
Orange Labs
Issy-les-Moulineaux
France

Yuanyuan ZHANG
CITI-INSA Lyon
INRIA
France

Wassim ZNAÏDI
CITI-INSA Lyon
INRIA
France

Index

A

ad hoc network, 12, 73-75, 77,
    83, 88, 96, 97
algebraic
    resolution, 6
    security, 106-111
ARQ, 30, 43, 44, 172

E, F

eavesdropper, 117-120
flow model, 266, 272, 273, 285
fountain codes, 28-45, 49, 59,
    61, 66, 67, 174

H

homomorphic
    ciphering, 120-126
    encryption, 123-128, 142
    MAC, 132, 134, 137
    signature, 119, 134-137

J

joint source-network, 236,
    249-251, 254-257

joint network-channel coding,
    245

L, M

Luby Transform (LT), 34
max-flow, 185, 236, 269,
    278-282
max-flow min-cut, 2-7, 270,
    271
multihop, 10, 11, 16, 24,
    42, 45, 49, 50, 55, 58,
    59, 66

O, P

optimization, 251, 255, 256,
    258, 268, 269, 284
P2P, 10, 21, 22, 24

R, S, T

random network coding, 8, 10,
    15-15, 20, 21, 119, 149-183,
    238, 258
security, 2, 99-102, 104-111,
    115, 117, 118, 120, 128, 130,
    132, 135, 142

sensor network, 43, 45, 47,
    49, 50, 55, 58, 61-67,
    112, 188, 251, 259
subspace code, 177, 242-245,
    256
TCP, 2, 10, 12, 19-21, 24,
    284

W, X

wireless sensor network (WSN),
    28, 29, 32, 41-44, 50, 64
wiretap network, 100, 102-105,
    112,
    type-II, 101, 102, 112
XOR network coding, 50